Disclaimer

The publisher of this book is by no way associated with the National Institute of Standards and Technology (NIST). The NIST did not publish this book. It was published by 50 page publications under the public domain license.

50 Page Publications.

Book Title: Electromagnetics Division: Programs, Activities, and Accomplishments

Book Author: Ronald B. Goldfarb

Book Abstract: The Electromagnetics Division is a critical national resource for a wide range of customers. U.S. industry is the primary customer both for the division's measurement services and for technical support on the test and measurement methodology necessary for research, product development, manufacturing, and international trade. The division represents the U.S. in international measurement intercomparisons and standards development related to radio-frequency and microwave technology, electromagnetic fields, and superconductors. The division also provides measurement services and expert technical support to other agencies of the federal government to support its programs in domestic and international commerce, national defense, transportation and communication, public health and safety, and law enforcement.

Citation: NIST Interagency/Internal Report (NISTIR)

Keyword: antennas;electromagnetic fields;electromagnetic measurements;electromagnetic properties of materials;emissions and immunity;magnetic recording;magnetic thin films and devices;magnetodynamics;nanoprobe imaging;noise;nonlinear device characterization;power;scattering parameters;superconductor electromechanical measurements;superconductor standards

NIST

National Institute of Standards and Technology

Technology Administration

U.S. Department of Commerce

NISTIR 7371

January 2007

Electronics and Electrical Engineering Laboratory

Electromagnetics Division

Programs, Activities, and Accomplishments

THE ELECTRONICS AND ELECTRICAL ENGINEERING LABORATORY

One of NIST's seven Measurement and Standards Laboratories, EEEL conducts research, provides measurement services, and helps set standards in support of: the fundamental electronic technologies of semiconductors, magnetics, and superconductors; information and communications technologies, such as fiber optics, photonics, microwaves, electronic displays, and electronics manufacturing supply chain collaboration; forensics and security measurement instrumentation; fundamental and practical physical standards and measurement services for electrical quantities; maintaining the quality and integrity of electrical power systems; and the development of nanoscale and microelectromechanical devices. EEEL provides support to law enforcement, corrections, and criminal justice agencies, including homeland security.

EEEL consists of four programmatic divisions and two matrix-managed offices:

Semiconductor Electronics Division

Optoelectronics Division

Quantum Electrical Metrology Division

Electromagnetics Division

Office of Microelectronics Programs

Office of Law Enforcement Standards

This document describes the technical programs of the Electromagnetics Division. Similar documents describing the other Divisions and Offices are available. Contact NIST/EEEL, 100 Bureau Drive, MS 8100, Gaithersburg, MD 20899-8100, telephone: (301) 975-2220, on the Web: http://www.eeel.nist.gov

On the cover (clockwise from upper left): extrapolation range for antenna measurements covering 50 to 110 gigahertz, portable system for measuring shielding of aircraft, vacuum chip bonder for making chip-scale alkali vapor cells, permittivity of fluids and foams, terahertz imaging system, spin-transfer oscillator measurements.

**ELECTRONICS AND ELECTRICAL
ENGINEERING LABORATORY**

ELECTROMAGNETICS
DIVISION

**PROGRAMS, ACTIVITIES, AND
ACCOMPLISHMENTS**

NISTIR 7371

January 2007

U.S. DEPARTMENT OF COMMERCE
Carlos M. Gutierrez, Secretary

Technology Administration
Robert Cresanti, Under Secretary of Commerce for Technology

National Institute of Standards and Technology
William Jeffrey, Director

Electromagnetics Division
National Institute of Standards and Technology
325 Broadway
Boulder, Colorado 80305-3328

Telephone: (303) 497-3131
Facsimile: (303) 497-3122
On the Web: http://www.boulder.nist.gov/div818/

Any mention of commercial products is for information only; it does not imply recommendation or endorsement by the National Institute of Standards and Technology nor does it imply that the products mentioned are necessarily the best available for the purpose.

Contents

Welcome iv
Introduction to Technical Programs 1
Electromagnetics Division Staff 3
Radio-Frequency Electronics Group
    Microwave Measurement Services 5
    Micro/Nano Electronics 11
    Electromagnetic Properties of Materials 16
Radio-Frequency Fields Group
    Antennas and Antenna Systems: Antenna Theory and Applications 20
    Antennas and Antenna Systems: Antenna Near-Field Measurements 22
    Antennas and Antenna Systems: Metrology for Radar Cross Section Systems 24
    Electromagnetic Compatibility: Reference Fields and Probes 26
    Electromagnetic Compatibility: Complex Fields 28
    Electromagnetic Compatibility: Time-Domain Fields 31
Magnetics Group
    Magneto-Mechanical Measurements for High Current Applications 33
    Standards for Superconductor and Magnetic Measurements 39
    Magnetodynamics 45
    Spin Electronics 49
    Magnetic Devices and Nanostructures 53
    Microsystems for Bio-Imaging and Metrology 59
Appendix A: Calibration Services 67
Appendix B: Postdoctoral Research Associateships 68
Appendix C: Prefixes for the International System of Units (SI) 81
Appendix D: Units for Magnetic Properties 82
Appendix E: Symbols for the Chemical Elements 83

WELCOME

The Electromagnetics Division is a critical national resource for a wide range of customers. U.S. industry is the primary customer both for the division's measurement services and for technical support of test and measurement methodology necessary for research, product development, manufacturing, and international trade. The division represents the U.S. in international measurement comparisons and standards development related to radio-frequency and microwave technology, electromagnetic fields, and superconductors. The division also provides measurement services and expert technical support to other agencies of the federal government in support of its programs in domestic and international commerce, national defense, transportation and communication, public health and safety, and law enforcement.

The division is organized into three groups. The groups' projects are led by senior technical staff with the assistance of physicists, engineers, technicians, and research associates, as well as graduate and undergraduate students.

The **Radio-Frequency Electronics Group** conducts theoretical and experimental research to develop basic metrology, special measurement techniques, and measurement standards necessary for advancing both conventional and microcircuit guided-wave technologies; for characterizing active and passive devices and networks; and for providing measurement services for scattering parameters, power, waveform, noise, material properties, and other basic quantities.

The **Radio-Frequency Fields Group** conducts theoretical and experimental research necessary for the accurate measurement of free-space electromagnetic field quantities; for the characterization of antennas, probes and antenna systems; for the development of effective methods for electromagnetic compatibility assessment; for the measurement of radar cross section and radiated noise; and for providing measurement services for essential parameters.

The **Magnetics Group** develops measurement technology for industries broadly concerned with magnetic information storage, biomagnetic imaging, and superconductor power, spanning the range from practical engineering to theoretical modeling. The group disseminates the results of its research through publications in refereed journals, presentations at conferences and workshops, and participation in standards organizations.

We hope that this collection of information will help you in understanding the work of the division and in making use of the technical capabilities and services that we provide for industry, government, and academia. We invite you to visit our Web site, http://www.boulder.nist.gov/div818. This site will provide you with more information on our projects and measurement related software, and reprints of our publications. Thank you for your interest in the Electromagnetics Division.

Michael Kelley, Chief, Electromagnetics Division

Ron Ginley, Acting Leader, Radio-Frequency Electronics Group

Perry Wilson, Leader, Radio-Frequency Fields Group

Ron Goldfarb, Leader, Magnetics Group

Introduction to Technical Programs

The division carries out a broad range of technical programs focused on the precise realization and measurement of physical quantities throughout the radio spectrum. Key directions include: (a) the development of reference standard artifacts, services and processes with which industry can maintain internationally recognized measurement traceability; (b) the advancement of technology through the development of new measurement techniques that are theoretically and experimentally sound as well as relevant and practical; (c) the assessment of total measurement uncertainties; and (d) the provision of expert technical support for national and international standards activities. We strive to perform leading-edge, high-quality research in metrology that is responsive to national needs. Division programs cover the following technical areas:

Microwave Measurement Services

Through the microwave measurement services, we establish, maintain, and disseminate the national standards for RF and microwave quantities. This provides the U.S scientific and industrial base with access to a measurement system that is readily available, reliable, accurate, reproducible, and internationally consistent. We provide a broad range of state-of-the-art RF and microwave calibration services, which include scattering parameters, attenuation, power, thermal noise, and waveform.

Micro/Nano Electronics

This program seeks to develop new metrology tools for next-generation electromagnetic applications in high-speed telecommunications, microelectronics, magnetic data storage, computing, and biomedical systems. These new tools are focused on higher operating frequencies, wider bandwidths, increased dynamic range, smaller length scale, and more complex materials, devices, and systems.

Electromagnetic Properties of Materials

The electromagnetic properties of materials program develops theory and methods for measuring the complex permittivities and permeabilities of materials at frequencies from RF through terahertz. Materials of interest include electronic substrates, thin films, liquids, biological specimens, and engineered materials. Characterization is performed of both bulk and micro- and nanoscale properties.

Antennas and Antenna Systems

The antennas and antenna systems program develops theory and techniques for measuring the gain, pattern, and polarization of advanced antennas; for measuring the gain and noise of large antenna systems; and for analyzing radar cross-section measurement systems.

Electromagnetic Compatibility

The electromagnetic compatibility program develops theory and methods for measuring electromagnetic field quantities and for characterizing the emissions and susceptibilities of electronic devices and systems, in both the frequency and time domains.

Superconductor Characterization and Standards

The program in superconductor characterization and standards develops measurement methods for the electric, magnetic, and mechanical properties of high-temperature and low-temperature superconductor wires and tapes for electric power applications.

MAGNETODYNAMICS, SPIN ELECTRONICS, AND MAGNETIC NANOSTRUCTURES

The program in magnetodynamics and spin electronics undertakes experimental research on fundamental aspects of magnetization switching, precession, and damping at the nanoscale for applications in magnetic information storage (such as magnetoresistive read heads, recording media, and magnetic random-access memory) and magnetic devices (such as oscillators driven by the transfer of quantum-mechanical electron spin angular momentum to magnetic films).

MAGNETIC RESONANCE AND MICROMECHANICAL SYSTEMS FOR BIOMAGNETIC IMAGING AND METROLOGY

This program is developing new metrology and standards for biomedical imaging based on magnetic resonance, magnetic nanoparticles, and low field sensing of biomagnetic fields. Micromechanical and magnetic resonance systems are being developed to characterize nanomagnetic systems and their use for measuring, detecting, and manipulating biomolecules for health and national security applications.

Electromagnetics Division Staff

Michael Kelley, Division Chief .... (303) 497-3131
Dennis Friday, Chief Scientist .... -3133
Cindy Kotary, Secretary .... -3132
Linda Derr, Administrative Officer .... -4202
Kate Remley, Technical Liaison .... -3652
Gerome Reeve, Research Associate .... -3557
Claude Weil, Research Associate .... -5305
Lindsey Vaughan, Student .... -4673

Radio-Frequency Electronics Group

Ron Ginley, Acting Group Leader .... (303) 497-3634
Susie Rivera, Secretary .... -5755

Microwave Measurement Services

Ron Ginley, Project Leader .... -3634
Robert Billinger .... -5737
Tom Crowley .... -4133
Puanani DeLara .... -3753
Denis LeGolvan .... -3210
Ann Monke .... -5249
Juanita Morgan .... -3015
James Randa .... -3150
David Walker .... -5490
Dylan Williams .... -3138

Micro/Nano Electronics

Mike Janezic, Project Leader .... -3656
Jim Baker-Jarvis .... -5490
Rob Billinger .... -5737
Jim Booth .... -7900
Pavel Kabos .... -3997
Nita Morgan .... -3015
James Randa .... -3150
Kate Remley .... -3652
Bill Riddle .... -5752
Dave Walker .... -5490
Dylan Williams .... -3138

Micro/Nano Electronics (continued)

Amanda Cox, Research Associate .... -4653
Eyal Gerecht, Research Associate .... -4199
Dazhen Gu, Research Associate .... -3939
Atif Imtiaz, Research Associate .... -4938
Arek Lewandowski,
    Research Associate .... -4665
Mu-Hong (Martin) Lin,
    Research Associate .... -4369
Hans Nembach, Research Associate .... -4966
Mitch Wallis, Research Associate .... -5089
Nathan Orloff, Graduate Student .... -4698

Electromagnetic Properties of Materials

Jim Baker-Jarvis, Project Leader .... -5490
Jim Booth .... -7900
Mike Janezic .... -3656
Pavel Kabos .... -3997
Bill Riddle .... -5752
Atif Imtiaz, Research Associate .... -4938
Mu-Hong (Martin) Lin,
    Research Associate .... -4369
Jordi Mateu, Research Associate .... -4974
Mitch Wallis, Research Associate .... -5089
Robin Powers, Student .... -7773

Radio-Frequency Fields Group

Perry Wilson, Group Leader ....(303) 497-3406
Ruth Marie Lyons, Secretary ....-3321

Antenna Theory and Applications

Michael Francis, Project Leader ....-5873
Lorant Muth ....-3603
Ronald Wittmann ....-3326
Randal Direen, Research Associate .....-5766

Antenna Near-Field Measurements

Katie MacReynolds, Project Leader ....-3471
Jeffrey Guerrieri ....-3863
Doug Tamura ....-3694
Carl Stubenrauch,
   Research Associate ....-3927

Reference Fields and Probes

Perry Wilson, Project Leader ....-3406
Dennis Camell ....-3214

Reference Fields and Probes (continued)

David Novotny ....-3168
Keith Masterson, Research Associate ..-3756

Complex Fields

Galen Koepke, Project Leader ....-5766
Christopher Holloway ....-6184
John Ladbury ....-5372
Jason Coder, Research Associate ....-4670
David Hill, Research Associate ....-3472
Ed Kuester, Research Associate ....-4312
William Young, Research Associate ....-4649

Time-Domain Fields

Robert Johnk, Project Leader ....-3737
Chriss Grosvenor ....-5958

Magnetics Group

Ron Goldfarb, Group Leader ....(303) 497-3650
Ruth Corwin, Secretary ....-5477

Magneto-Mechanical Measurements for High Current Applications

Jack Ekin, Project Leader ....-5448
Cam Clickner ....-5441
Najib Cheggour, Research Associate ...-3815
Danko van der Laan,
   Research Associate ....-4702

Standards for Superconductor and Magnetic Measurements

Loren Goodrich, Project Leader ....-3143
Ted Stauffer ....-3777

Magnetodynamics

Tom Silva, Project Leader ....-7826
Tony Kos ....-5333
Mark Hoefer, Research Associate ....-4282

Spin Electronics

Bill Rippard, Project Leader ....-3882
Matt Pufall, Research Associate ....-5206
Mike Schneider, Research Associate ...-4203

Magnetic Devices and Nanostructures

Stephen Russek, Project Leader ....-5097
Brant Cage, Research Associate ....-4224
Justin Shaw, Research Associate ....-4421

Microsystems for Bio-Imaging and Metrology

John Moreland, Project Leader ....-3641
Li-Anne Liew, Research Associate ......-4197
Gary Zabow, Research Associate ....-4657
Wendy Krauser, Graduate Student ....-4350
Dan Porpora, Graduate Student ....-4458

Microwave Measurement Services

Goals

The goal of the Microwave Measurement Service Project is to ensure the availability for the U.S. scientific and industrial base of a measurement system for radio frequency (**RF**) and microwave quantities that is reliable, accurate, reproducible, traceable to the International System of Units (**SI**), and internationally consistent. We do this by developing and maintaining the U.S. national measurement standards for RF and microwave quantities, providing a wide range of state-of-the-art calibration services, and developing new measurement methods and verification techniques to improve the quantitative measurement of microwave quantities.

Technical Contacts:
Ron Ginley
Tom Crowley
Dave Walker

Staff-Years (FY 2006):
3.7 professionals
4.1 technicians

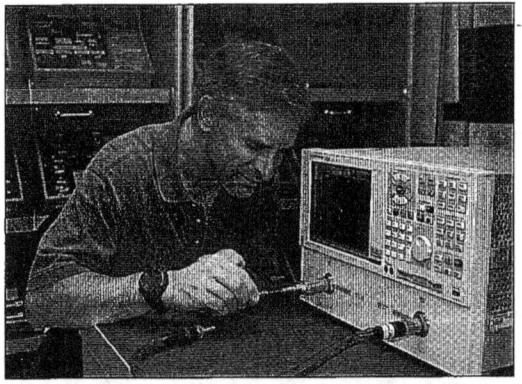

Setting up experimental vector network analyzer calibration using 1.5 millimeter airline standard.

Customer Needs

The customers who use our services span a large part of the electronics industry. They include aeronautics and communication companies, instrument manufacturers, and other government agencies. Additionally we make measurements for many internal programs in areas such as antennas, optoelectronics, and electromagnetic properties of materials. Our services provide the fundamental microwave properties that customers rely on to establish the critical factors in design and performance of RF and microwave equipment. Our customers also establish traceability to the SI through our measurements. Economic gains are realized through improvements in accuracy. The verification of calibrations and measurement processes on commonly used microwave measurement systems is of paramount importance to our customers. We support this through our measurements and the techniques that we make available.

Microwave metrology is expanding in many different directions. There is a constant push to use higher frequencies. Signals are becoming much more complex and include modulation effects, multiport/differential signals, complex waveforms, and other unusual signal schemes. On-wafer measurements are in greater demand. Improved means for the dissemination of our services are also necessary. These new requirements are dictated by the needs of the telecommunication and computing communities; 100 gigabit per second data rates will require 400 gigahertz support as well as modulated power, waveform analysis and other signal scheme support. Optoelectronic applications need scattering parameter (***s*-parameter**) and power measurement support above 50 gigahertz in 1.85 millimeter connectors. Molecular resonance measurements for chemical identification will need precision measurement support in the 500 to 700 gigahertz range. Remote sensing will require measurements of unprecedented accuracy. New imaging systems will require support in many different microwave parameter areas.

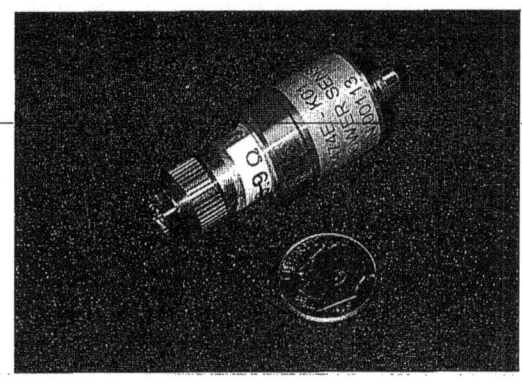

A 2.4 millimeter thin-film primary power transfer standard.

TECHNICAL STRATEGY

We provide a large range of state-of-the-art microwave measurements and standards. We will continue to maintain the primary national standards in thermal noise, s-parameters, power, and waveform metrology, and to offer measurement services that enable customers to achieve traceability to those standards and to verify their own measurements.

Maintaining and delivering these services is a major task. The systems and standards for the services are generally designed and built at NIST and are not commercially available. There is not a large market for primary standards and, therefore, companies have no economic incentive to develop them. To cover all parameters requires many different systems and standards. The systems are aging, need more maintenance time, and are very costly to replace. The primary standards that support the services are also aging. In many cases the present standards are either nonreplaceable, or the technical expertise to recreate them has been lost, or the parts of the standards obtained from commercial sources are no longer available (for example, the WR15 and WR10 thermistor standards that have been used are no longer made by any manufacturer).

In part, our strategy for moving the measurement services into the future is to develop improved methods for delivering our services. This will include alleviating the stress on measurement systems and standards, new methods of supporting our customer needs, and more interlaboratory comparisons.

SCATTERING PARAMETERS

S-parameter measurements are required for accurate measurements in all of the other microwave disciplines. An example is the mismatch correction for power measurements that is calculated from reflection coefficients of various parts of the system and devices. The accuracy of s-parameter measurements is directly related to the calibration of the vector network analyzer (**VNA**) used for the measurements. The calibration method and standards can be chosen to match the end use. It is important to educate the users of the modern network analyzers about the choices of standards and calibration methods and to develop tools that will enable them to make reliable, accurate measurements.

There is an increasing demand for s-parameter measurements. This is particularly evident for measurements above 50 gigahertz. Higher frequencies and smaller waveguide and connector sizes are starting to be used routinely. We are going to address these needs by adding 1.85 millimeter and 1.0 millimeter connector size capabilities to our measurement services in the near term and look at supporting small waveguide sizes (frequency coverage up to at least 500 gigahertz) in the longer term.

There is general agreement among the principal users and makers of VNA systems that much still remains to be understood about VNA calibrations and measurements. This is true for both traditional measurement areas and emerging areas, which include electronic calibration units, multiport, and differential measurements. We plan to take a very active role in developing VNA calibration and measurement theory and techniques for these areas. We intend to pursue these VNA techniques to higher frequency ranges, up to approximately 500 gigahertz.

Verification of VNA calibrations is very important, and the current verification process is not sufficient. We will try to improve the verification process through several different methods. The Verify and Calkit software we have developed, which compares the contents on verification and calibration disks to measurements made based on LRL calibrations, will be made available to the public. Our measurement comparison kit program will be expanded. Finally services will be developed to support ongoing interlaboratory comparisons. These will aid not only in system verification, but also in proficiency testing for accreditation.

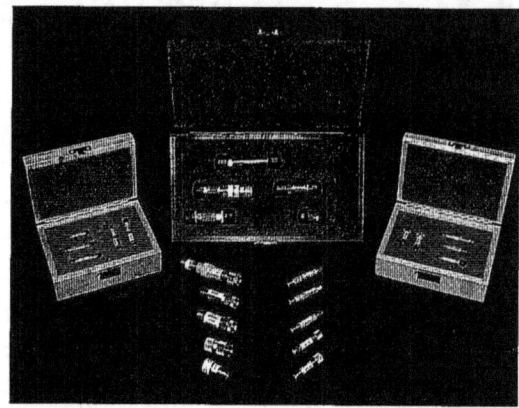

S-parameter measurement comparison kits for Type-N 7 millimeter, 3.5 millimeter, 2.92 millimeter and 2.4 millimeter coaxial connectors.

NOISE

We have recently added the capability to perform noise-parameter measurements on low-noise amplifiers, and we plan to develop mechanisms, such as verification methods, to support such measurements in industry. In conjunction with our on-wafer noise efforts (see the Micro/Nano Electronics Project), we are working to improve on-wafer measurement methods for noise parameters of transistors, and we will also develop methods of supporting those measurements in industry.

New microwave remote-sensing radiometers are designed for unprecedented accuracy. In order to verify (and possibly to even achieve) that accuracy, a stable, accurate reference standard is required. We have proposed development of national standards for microwave brightness temperature in frequency bands of interest. The standards would be traceable to the NIST primary noise standards. We are also working on methods to characterize calibration targets commonly used in microwave radiometry. Additionally we are automating our noise temperature systems to greatly increase their efficiency.

Power

For a number of years, power measurements above 50 gigahertz were based on calorimetric and six-port reflectometer measurements using rectangular waveguides. NIST's internal power standards were characterized in the calorimeters and the measurements transferred to customer devices by use of the six-port systems. The NIST primary transfer standards were modified commercial power sensors, and the calorimeters were designed specifically for these standards. Measurement services were available in 1 gigahertz steps from 50 to 75 gigahertz and from 92 to 96 gigahertz.

Improvements in our standards are needed for a number of reasons. Frequency coverage in 1 gigahertz steps is not adequate for characterizing broadband digital devices such as optoelectronics components that operate at 40 gigabits per second. There is increasing use of frequencies above 75 gigahertz that were not previously measured. The modified commercial power sensors that NIST used as transfer standards are no longer reliable and cannot be replaced.

In order to address these problems, new calorimeters are being developed so that a wider set of transfer standards can be used. Faster transfer measurements are being developed so that denser frequency coverage can be readily obtained. We have obtained a new synthesizer (up to 67 gigahertz) and a backward-wave oscillator (50 to 110 gigahertz) source that have replaced the manually tuned Gunn diode oscillators for most measurements. A direct comparison system that can evaluate a customer device at about 50 frequencies per day has been developed for WR-15 (50 to 75 gigahertz) and is being used for customer calibrations. A direct comparison system for WR-10 (75 to 110 gigahertz) has been constructed but not yet evaluated. New calorimeters are being designed for both WR-10 and WR-15. They will accommodate a wider variety of internal standards than the present calorimeters. Future plans include the extension of the direct comparison measurements to 1.85 millimeter coaxial connectors that will allow measurements from DC to 65 gigahertz with a single connector. These measurements will be traceable to the WR-15 and 2.4 millimeter calorimetric primary standards.

RF power measurements have traditionally been traceable to DC power measurements through RF/DC substitution techniques. An alternative approach is to measure the field strength of microwaves through their effect on the quantum state of atoms. In this measurement, a group of atoms is created in a single quantum state. They are then exposed to microwaves at a frequency that corresponds to the energy difference between this state and a second quantum state. The atoms will oscillate between the two states at a frequency that is proportional to the field strength. This process is known as a Rabi oscillation. By measuring the number of atoms in each state, the field strength can be determined. A proof-of-concept experiment was conducted in collaboration with the Physics Laboratory. The next stage in this work will be an experiment that accurately compares a traditional measurement with the quantum measurement.

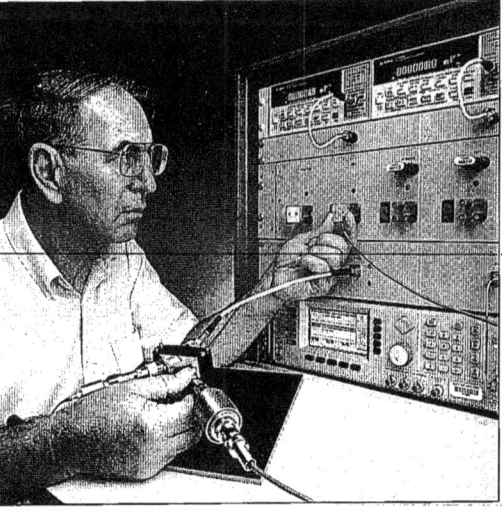

Calibration measurements on the direct comparison power system.

Waveform

We are developing calibration procedures for today's high-performance electrical probes for on-wafer measurement. We are also laying the foundations for 200 gigahertz to 400 gigahertz calibrations for tomorrow's on-wafer probes. We are also developing techniques for performing non-invasive high-impedance on-wafer waveform measurements for characterization of signal integrity in digital silicon integrated circuits (**ICs**) and other small circuits. This effort is particularly important for the development of electrical metrology for nanoscale devices, which, due to their small sizes, have extremely high electrical impedances.

Our plan is to electrically characterize an active high-impedance probe with our existing VNA calibration methods. We will then characterize the same probe on our 200 gigahertz bandwidth electro-optic sampling (**EOS**) system. This will lay the groundwork for very high-speed, on-wafer calibrations for digital IC and nanoelectronics. We will develop joint time-domain/frequency-domain uncertainty analysis for coaxial photodiode pulse sources. The calibration and uncertainty representation will include imperfections in the electro-optic sampling system and electrical mismatch corrections, and will be suitable for calibrating oscilloscopes with coaxial ports in either the time or frequency domains to 110 gigahertz. We will develop pulse sources with 400 gigahertz bandwidth and, based on these sources, develop on-wafer waveform characterization ability to 400 gigahertz. We will apply high-speed waveform metrology to microwave problems, including the characterization of electrical phase standards, microwave sources, and microwave mixers.

Accomplishments

- **Low Noise Amplifiers** — Low-noise amplifiers (**LNAs**) are used in a variety of applications involving detection and processing of low-level signals; noise parameters are the critical characteristics of LNAs. We performed measurements of the noise parameters of an LNA for an informal comparison with another National Measurement Institute (**NMI**) in order to verify international agreement on such measurements.

- **Microwave Brightness-Temperature Standards** — New microwave remote-sensing radiometers for weather, climate monitoring, and other applications are designed for unprecedented accuracy. In order to verify (and possibly to even achieve) that accuracy, a stable, accurate reference standard is required. We proposed and documented a plan for development of national microwave brightness-temperature standards, traceable to fundamental noise standards.

- **Calibration Targets** — Heated calibration targets are commonly used as standards for microwave remote sensing radiometers, but important characteristics of such targets are not well measured. We performed preliminary measurements of the electromagnetic properties of materials commonly used in calibration targets with the Electromagnetic Properties of Materials Project and also performed infrared imaging of thermal gradients in a target borrowed from NASA with the Physics Laboratory. With the Radio-Frequency Fields Group, we performed a preliminary study of near-field effects on the calibration of microwave radiometers when the calibration target is in the near field of the radiometer antenna. It is current practice to ignore such effects; however, this study indicated that they could be significant.

- **International Comparisons** — Under the Mutual Recognition Agreement (**MRA**) of NMIs worldwide, measurement comparisons among NMIs are performed on key quantities to assure international harmonization of standards and of measurements for commercial regulatory purposes. We participated in two completed Consultative Committee for Electricity and Magnetism (**CCEM**) Key Comparisons (**KCs**) in noise, one KC in s-parameters, one KC in power, and one supplementary comparison in power, for which we were the pilot lab.

- **WR-15 and WR-10 Waveguide Systems** — Waveguide services above 50 gigahertz are being revitalized to compensate for difficulties with existing transfer standards and to improve the traceability for high frequency signals used in digital communications. A direct-comparison system for WR-15 waveguide has been implemented and is being used for customer calibrations. A WR-10 waveguide system has been constructed. These systems greatly reduce the time required for a measurement. New WR-15 and WR-10 calorimeters were designed and tested. After initial tests, it was decided to modify the new design to improve their performance.

- **Microwave Field Strength** — Basic research testing a fundamentally new method for measuring microwave signal strength was conducted in collaboration with the Physics Laboratory. The RF magnetic field strength in a cavity was measured by observing the oscillation of cesium atoms between two quantum states. The proof-of-principle experiment demonstrated a rough agreement between the new technique and traditional microwave power measurements.

- **Power Standards at Low Frequency** — A new coaxial direct comparison system has been built to cover the frequency range of 100 kilohertz to 18 gigahertz. This system extends the lower range of our direct comparison systems from 50 megahertz down to 100 kilohertz. This new system gives us much greater efficiency in measuring power standards. Before, two independent systems had to be used. There was no significant change in uncertainties.

- **Software** — The Verify and Calkit software that supports the Agilent 8753 VNA has been completed and delivered to the U.S. Air Force. The Air Force will be using this to verify in-house their inventory of Agilent 8753 verification and calibration kits at an estimated cost savings of approximately $100,000 per year.

CALIBRATIONS

- We provide a wide range of state-of-the-art calibration services for fundamental microwave quantities, including scattering parameter, power, thermal noise, and waveform.

- For scattering parameters, we provide calibrations of one- and two-port devices, in a variety of waveguides and coaxial connector sizes. For coax, we provide measurements from 10 megahertz to 50 gigahertz. In waveguide, we cover the range from 8.2 to 110 gigahertz and 92 to 96 gigahertz.

- For microwave power, we provide effective efficiency calibrations of thermistor and thermoelectric detectors in coax from 100 kilohertz to 50 gigahertz. We provide calibration in waveguide for thermistor detectors from 8.2 to 75 gigahertz.

- For thermal noise, we provide calibration of coaxial noise standards at 30 megahertz, 60 megahertz, and from 1.0 to 50 gigahertz. Waveguide standards are calibrated from 8.2 to 65 gigahertz.

- For waveform, in collaboration with the Optoelectronics Division, we provide calibration of photodetectors and oscilloscopes for microwave signal characterization. We provide measurements in coaxial media at frequencies up to 110 gigahertz and on-wafer up to 200 gigahertz.

SHORT COURSES

- We organized the annual Automatic RF Techniques Group (**ARFTG**)/NIST Short Course on Microwave Measurements in Washington, DC, in November 2005. The course is a great way to bring measurement theory and techniques to many people (attendance is usually around 40) at one time. During the course we are able to teach the attendees the latest methods for making the most accurate microwave measurements. Presentations were given by Jim Randa on "Thermal Noise Measurements," Ron Ginley on "RF Connectors and Transmission Lines" and "VNA Uncertainties," and Tom Crowley on "Microwave Power Measurements."

- Dave Walker gave lectures at the Asia-Pacific Microwave Measurements Training Course in Christchurch, New Zealand, in April 2006, and in Auckland, New Zealand, in May 2006: "Introduction to Microwave Power Measurements," "Microwave Power Measurements in Digital Communications Systems," and "Thermal Noise Measurements."

RECENT PUBLICATIONS

T. P. Crowley, J. Miall, J. P. M. de Vreede, J. Furrer, A. Michaud, E. Dressler, T. Zhang, K. Shimaoka, and J. H. Kim, "CCEM Supplementary Comparison: RF Power Measurements with 2.4 mm Connectors," CCEM.RF-S1.CL (GTRF/02-03)," *Metrologia (Technical Supplement)* **43**, 01007 (October 2006).

C. Eiø, D. Adamson, J. Randa, D. Allal, and R. Uzdin, "CCEM Key Comparison: Noise in 50 Ω Coaxial Line at Frequencies up to 1 GHz," CCEM.RF-K18.CL (GT-RF/00-1), *Metrologia (Technical Supplement)* **43**, 01004 (October 2006).

J. Randa, A. E. Cox, and D. K. Walker, "Proposal for Development of a National Microwave Brightness-Temperature Standard," *Proc. SPIE* **6301**, 630105 (September 2006).

J. Randa, A. E. Cox, and D. K. Walker, "Proposed Development of a National Standard for Microwave Brightness Temperature," Int. Geoscience and Remote Sensing Symp. (IGARSS-2006), Conference Digest, Denver, CO (August 2006).

A. E. Cox and M. D. Janezic, "Preliminary Studies of Electromagnetic Properties of Microwave Absorbing Materials used in Calibration Targets," Digest Int. Geoscience and Remote Sensing Symp. (IGARSS-2006), Denver, CO (August 2006).

WR 15 six-port reflectometer head.

A. E. Cox, J. J. O'Connell, and J. P. Rice, "Initial Results from the Infrared Calibration and Infrared Imaging of a Microwave Calibration Target," Digest Int. Geoscience and Remote Sensing Symp. (IGARSS-2006), Denver, CO (August 2006).

D. Williams, H. Khenissi, F. Ndagijimana, K. A. Remley, J. Dunsmore, P. D. Hale, J. Wang, and T. S. Clement, "Microwave-Mixer Measurement with a Sampling Oscilloscope," *IEEE Microwave Theory Tech.* **54**, 1210-1217 (March 2006).

R. Ginley, "A Direct Comparison System for Measuring Low Frequency Power (100 kHz to 18 GHz)," Proc. Meas. Sci. Conf. (MSC) 2006, Anaheim, CA (March 2006).

D. Williams, A. Lewandowski, T. S. Clement, C. M. Wang, P. D. Hale, J. M. Morgan, D. A. Keenan, A. K. Dienstfrey, "Covariance-Based Uncertainty Analysis of the NIST Electro-Optic Sampling System," *IEEE Trans. Microwave Theory Tech.* **54**, 481-491 (January 2006).

D. Williams, P. D. Hale, and T. S. Clement, "Calibrated 200 GHz Waveform Measurement," *IEEE Photon. Tech. Lett.*, **53**, 1384-1389 (April 2005).

J. Randa, D. K. Walker, A. E. Cox, and R. L. Billinger, "Errors Resulting From the Reflectivity of Calibration Targets," *IEEE Trans. Geosci. Remote Sens.* **43**, 50-58 (January 2005).

T. P. Crowley, E. A. Donley, and T. P. Heavner, "Quantum-Based Microwave Power Measurements: Proof-of-Concept Experiment," *Rev. Sci. Instrum.* **75**, 2575-2580 (August 2004).

Micro/Nano Electronics

Goals

The next generation of high-frequency electromagnetic applications will require higher operating frequencies, wider bandwidths, increased dynamic range, and a substantial reduction in device dimensions. To meet these challenges, the Micro/Nano Electronics Project develops new measurement techniques, standards, and instrumentation to satisfy the growing metrology needs of the high-speed telecommunications, microelectronics, magnetic storage, and computing industries.

Customer Needs

The full impact of nanotechnology research may be decades away, but one current critical need, according to the National Nanotechnology Initiative (**NNI**), is the creation and development of the fundamental metrology necessary to characterize nanoscale electrical and magnetic devices. An additional complexity is that future applications of both microwave and nanotechnology will involve the transmission and detection of broad-bandwidth signals having complicated modulation schemes at frequencies ranging from the microwave region to the terahertz region. Therefore, development of nanotechnology metrology will require synergy between research areas as diverse as nanoelectronics, telecommunications, microwave metrology, and materials science.

Technical Strategy

To support the future needs of the microelectronics and emerging nanoelectronics industries, both the International Technology Roadmap for Semiconductors (**ITRS**) and the NNI outline many of the metrology needs that must be addressed for the next generation of microelectronic and nanoelectronic devices. In both of these industries, metrology must be developed for characterization of smaller devices that carry complicated, modulated signals at frequencies that extend into and beyond the microwave range. This project's expertise in microwave and terahertz metrology, high-speed waveform metrology, measurements for telecommunication systems, and high-frequency nanodevice characterization allows it to tackle some of the major metrology issues facing the microelectronics and nanoelectronics industries.

This project has played an important role in developing the fundamental tools for performing on-wafer microwave measurements. Building on this expertise, we are extending traditional on-wafer techniques to develop the necessary metrology for supporting the telecommunications and emerging nanoelectronics industry. One such extension is in the area of high-speed waveform metrology. As data rates of optical links exceed 40 gigabits per second, and emerging digital circuits operate with clock rates over 100 gigahertz, electrical measurements with conventional techniques no longer work; they are invasive and limited in frequency. To meet this need, we have developed a fully calibrated electro-optic sampling system for characterizing photodetectors and calibrating oscilloscopes for microwave signal characterization. This has enabled us to develop new tools, such as electro-optic on-wafer scattering and waveform measurements up to 200 gigahertz and modulated microwave signal and coaxial signal-source characterization to 400 gigahertz. With these tools, we are providing the fundamental metrology that will enable development of the next generation of high-frequency oscilloscopes, microwave transition analyzers (**MTAs**), and related instruments.

High-frequency characterization of nanodevices is another focus of research in this project. Development of atomic force microscopes (**AFMs**), scanning tunneling microscopes (**STMs**), and high-impedance and near-field evanescent probing systems that operate at high frequencies allows for characterization of the electromagnetic fields, materials, and nanodevice electrical properties at frequencies in the microwave and millimeter-wave range with submicrometer resolution.

Another extension of the on-wafer microwave metrology work is in the area of on-wafer noise characterization. Here, we are developing new measurement techniques that can be used to characterize the on-wafer noise properties of complementary metal oxide semiconductor (**CMOS**) devices, which can be used not only to improve device performance but to assist in physical model development. In addition to the measurements performed on-wafer, we are also developing new techniques for noise characterization at terahertz frequencies, which will provide traceability to a critical parameter of active terahertz devices.

In addition to the on-wafer activities, the project also develops measurement methods and calibrations for coaxial instrumentation capable of accurately characterizing RF-based systems that transmit complex modulated signals. These include

Technical Contacts:
Mike Janezic
Pavel Kabos
Jim Randa
Kate Remley
Dylan Williams

Staff-Years (FY 2006):
3.0 professionals
0.7 technician

next-generation and broadband wireless system technologies, as well as broadband systems utilized by first responders to emergencies. The project is also involved in development of test signals that will provide easier calibrations for wireless system measurements.

Besides the characterization of noise at terahertz frequencies, we are also focusing on applications of passive terahertz imaging. Terahertz radiation can penetrate clothing and, to some extent, can also penetrate biological materials. Because of their shorter wavelengths they offer higher spatial resolution than microwaves or millimeter-waves. This project focuses on developing the critical metrology necessary to characterize newly developed terahertz devices and systems that are coming to market.

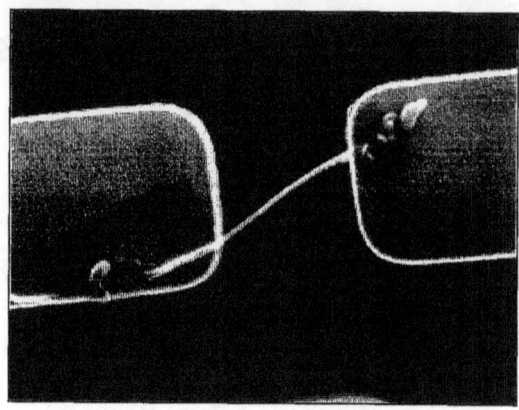

A multi-wall carbon nanotube welded across a gap in the center conductor of a coplanar waveguide.

Accomplishments

■ **High-Frequency Nanodevice Metrology** — The primary focus is the development of metrology for high-frequency scanned-probe microscopy and micro-electromechanical systems (**MEMS**) with submicrometer resolution. Specifically, this includes the areas of quantitative high-frequency imaging, measurement of electromagnetic field components, and characterization of materials that are incorporated into nanoscale electrical and magnetic devices.

In order to investigate the high-frequency response of nanoscale structures such as carbon nanotubes and Si-based nanowires, we constructed a universal scanning probe station that can operate as a RF-STM probe, near-field scanning probe, AFM probe, or a high-impedance, noncontacting probe with nanometer spatial positioning at frequencies up to 50 gigahertz.

For more accurate measurements of electromagnetic field components, we designed a three-layer SiN/SiO/SiN cantilever probe for high-frequency calorimetric field imaging and showed a demonstrable improvement in performance and yield over traditional two-layer micro-electromechanical system (**MEMS**) calorimetric probes.

Investigating the Einstein–de Haas effect, we developed a MEMS-based approach to measuring the magneto-mechanical ratio g', a critical parameter for the development of materials for spin electronics and magnetic data storage.

We demonstrated that calibrated on-wafer techniques can be employed to characterize the broadband frequency response of high-impedance devices such as multiwalled carbon nanotubes as well as quantify the effect of contact resistances.

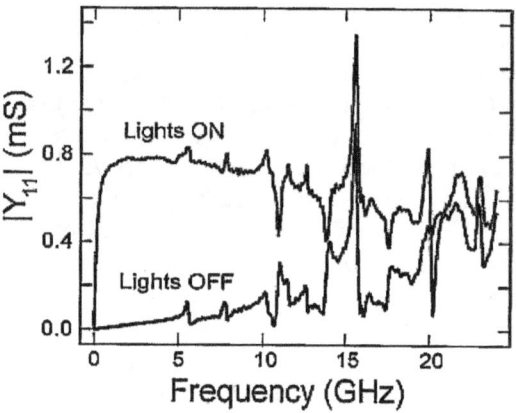

Admittance (which includes conductance) in millisiemens as a function of frequency in gigahertz at the left contact of the carbon nanotube. Data were taken with illumination of the nanotube on and off. The admittance is calculated from scattering parameters measured with a network analyzer.

■ **High-Speed Waveform Metrology** — Sales in the high-speed telecommunications market are based on the ability to deliver high-speed, reliable, and interoperable communication systems with greater information-carrying capacity and at a lower cost. Integrated digital and microwave electronics now require bandwidths in excess of 110 gigahertz, larger than can be supported by current coaxial connectors. To meet these demanding needs, we are developing both 110 gigahertz coaxial waveform metrology, as well as on-wafer metrology to even higher frequencies. Our current focus includes design of a 400 gigahertz opto-electronic source, integration of this source into a microwave probe, and construction of an electro-optic measurement system to characterize the bandwidth of this source.

We developed a coaxial waveform measurement service that is traceable to our electro-optic sampling system. It includes point-by-point uncertainties and correction for mismatches, and is traceable to 110 gigahertz. We demonstrated the ability to generate and accurately characterize pulses on-wafer with the NIST electro-optic sampling system up to 200 gigahertz, important for characterizing 40 gigabit and higher communications systems. We developed new measurements for characterizing microwave mixers and modulated signal sources. We also developed a new 110 gigahertz, 1.0 millimeter coaxial scattering-parameter calibration and covariance-based uncertainty analysis.

■ **Thermal Noise Characterization at Terahertz Frequencies** — Applications of electromagnetic measurements at terahertz frequencies are poised for explosive growth, but lack of fundamental standards is a serious impediment. Traceability to NIST noise standards at terahertz frequencies would allow comparison of different measurements and meaningful comparison of performance of components and systems.

For characterization of cryogenic amplifiers, we developed a method for accurate noise figure measurements. We measured effective input noise temperatures in the 1 to 12 gigahertz range, with results as low as 2.3 kelvins with a standard uncertainty of 0.3 kelvin, corresponding to a noise figure of 0.034 decibels ± 0.004 decibels. We developed and demonstrated a terahertz receiver with a quasi-optical adapter for coupling incident radiation into a receiver. The receiver is based on heterodyne detection with a hot-electron-bolometer (**HEB**) mixer. To house the receiver portion of the radiometer, we purchased and instrumented a cryocooler.

■ **Passive Spectral Terahertz Imaging** — Much interest exists in applications for imaging and spectroscopy at terahertz frequencies. We developed a two-dimensional scanning passive terahertz imaging system based on a phonon-cooled quasi-optically coupled HEB mixer that is integrated with an InP monolithic microwave integrated circuit (**MMIC**) intermediate frequency (**IF**) low-noise amplifier. We demonstrated the ability of a scanning passive terahertz imaging system to obtain full two-dimensional images of various objects by scanning the target with a flat mirror mounted on a computer-controlled elevation/azimuth translator. The overall sensitivity, defined in terms of noise equivalent delta temperature (**NEΔT**), is better than 0.5 kelvin.

■ **Metrology for Wireless Systems** — We are developing test methods to characterize signal degradation, including attenuation and phase distortion, in complex environments including large buildings and basements. In particular, we are focusing on development of wideband measurement methods in the new 4.9 gigahertz public safety radio band for transmission of voice, data, images, and video.

To improve measurements of wideband, modulated signals, we developed a method to extend the measurement bandwidth of narrowband vector receivers. We developed impedance mismatch correction techniques for vector signal generators in their large-signal operating state. In one recent application, we investigated the viability of using RF modulated-signal measurements order to identify nodes in a wireless local-area network. We performed measurements of wideband, modulated signals in typical first-responder environments.

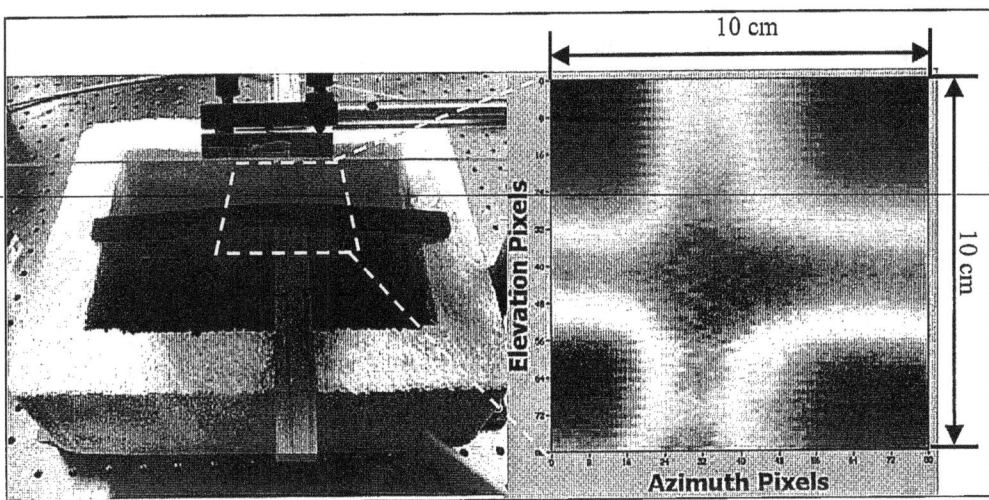

Photograph (left) and 0.850 terahertz image (right) of two objects at room temperature suspended over an absorber immersed in liquid nitrogen. The terahertz image covers a temperature range of 200 kelvins.

■ **On-Wafer Noise Measurements** — As transistor technology evolves to smaller, lower-noise devices used at higher frequencies, measuring their noise characteristics becomes increasingly difficult. The difficulties increase due to the very low noise levels that must be measured and to the very high reflections exhibited by such devices. We are developing accurate measurement techniques that can be used to evaluate and improve physical models, as well as design and predict the performance of systems that employ transistors.

In order to characterize poorly matched transistors, we extended and improved a Monte Carlo program used for amplifier noise parameter uncertainty analysis and used it to perform simulations comparing different measurement strategies. We developed the capability to measure noise parameter of on-wafer transistors or low-noise amplifiers. We performed a measurement comparison with two companies of on-wafer noise parameters of CMOS transistors with 0.13 micrometer gate lengths.

To better determine the optimal input source reflection coefficient, we developed an improved on-wafer noise-parameter measurement method.

WORKSHOPS

To disseminate newly developed on-wafer noise measurements, we organized a joint International Microwave Symposium (**IMS**)/IEEE Radio Frequency Integrated Circuits (**RFIC**) full-day workshop on "Noise Measurement and Modeling for CMOS" in San Francisco, CA, in June 2006.

New measurement techniques useful to wireless engineers were presented in two full-day workshops at the IMS conference organized by project members. One was on the topic "Memory Effects in Power Amplifiers," while the other was on "Techniques and Applications of Wireless Sensor Networks."

RECENT PUBLICATIONS

S. Lee, T. M. Wallis, J. Moreland, P. Kabos, and Y. C. Lee, "Asymmetric Dielectric Trilayer Cantilever Probe for Calorimetric High Frequency Field Imaging," *IEEE J. Microelectromech. Syst.* (in press).

E. Gerecht, D. Gu, F. Rodriguez-Morales, and K. S. Yngvesson, "Imaging and Spectroscopy at Terahertz Frequencies Using Hot Electron Bolometer Technology," Proc. SPIE Optics East, Boston, MA, October 2006 (in press).

T. M. Wallis, J. Moreland, and P. Kabos, "Einstein–de Haas Effect in a NiFe Film Deposited on a Microcantilever," *Appl. Phys. Lett.* **89**, 122502 (September 2006).

E. Gerecht, D. Gu, J. Randa, D. Walker, and R. Billinger, "Terahertz Radiometer Design for Traceable Noise-Temperature Measurements," Conf. Precision Electromagnetic Meas. (CPEM) Digest, Torino, Italy, July 2006, pp. 660-661.

D. K. Walker and J. Randa, "On-Wafer Noise-Parameter Measurements at NIST," Conf. Precision Electromagnetic Meas. (CPEM) Digest, Torino, Italy, July 2006, pp. 656-657.

F. Rodriguez-Morales, K. S. Yngvesson, R. Zannoni, E. Gerecht, D. Gu, X. Zhao, N. Wadefalk, and J. Nicholson, "Development of Integrated HEB/MMIC Receivers for Near-Range Terahertz Imaging," *IEEE Trans. Microwave Theory Tech.* **54**, 2301-2311 (June 2006).

J. Randa, T. McKay, S. L. Sweeney, D. K. Walker, L. Wagner, D. R. Greenberg, J. Tao, and G.A. Rezvani, "Reverse Noise Measurement and Use in Device Characterization," 2006 IEEE Radio Frequency Integrated Circuits (RFIC) Symp. Digest, San Francisco, pp. 345-348 (June 2006).

S. Lee, Y. C. Lee, T. M. Wallis, J. Moreland, and P. Kabos, "Near-Field Imaging of High-Frequency Magnetic Fields with Calorimetric Cantilever Probes," *J. Appl. Phys.* **99**, 08H306 (April 2006).

D. F. Williams, H. Khenissi, F. Ndagijimana, K. A. Remley, J. P. Dunsmore, P. D. Hale, C. M. Wang, and T. S. Clement, "Sampling-Oscilloscope Measurement of a Microwave Mixer with Single-Digit Phase Accuracy," *IEEE Trans. Microwave Theory Tech.* **53**, 1210-1217 (March 2006).

J. Randa, E. Gerecht, D. Gu, and R. L. Billinger, "Precision Measurement Method for Cryogenic Amplifier Noise Temperatures Below 5 K," *IEEE Trans. Microwave Theory Tech.* **54**, 1180-1189 (March 2006).

K. A. Remley, M. Rütschlin, D. F. Williams, R. T. Johnk, G. Koepke, C. L. Holloway, M. Worrell, and A. MacFarlane, "Radio Communications for Emergency Responders in Large Public Buildings: Comparing Analog and Digital Modulation," Proc. 2006 Intl. Symp. Advanced Radio Tech. (ISART), Boulder, CO, pp. 79-82 (March 2006).

D. F. Williams, A. Lewandowski, T. S. Clement, C. M. Wang, P. D. Hale, J. M. Morgan, D. Keenan, and A. Dienstfrey, "Covariance-Based Uncertainty Analysis of the NIST Electro-Optic Sampling System," *IEEE Trans. Microwave Theory Tech.* **54**, 481-491 (January 2006).

J. R. Baker-Jarvis, "Time-Dependent Entropy Evolution in Microscopic and Macroscopic Electromagnetic Relaxation," *Phys. Rev. E* **72**, 066613 (December 2005).

K. A. Remley, C. A. Grosvenor, R. T. Johnk, D. R. Novotny, P. D. Hale, M. D. McKinley, A. Karygiannis, and E. Antonakakis, "Electromagnetic Signatures of WLAN Cards and Network Security," Proc. IEEE Int. Symp. Signal Processing and Information Technology, Athens, Greece, December 2005, pp. 484-488.

K. A. Remley, P. D. Hale, D. I. Bergman, and D. Keenan, "Comparison of Multisine Measurements from Instrumentation Capable of Nonlinear System Characterization," 66th ARFTG Microwave Measurements Conf. Digest, Washington, DC, December 2005, pp. 34-43.

J. Randa, S. L. Sweeney, T. McKay, D. K. Walker, D. R. Greenberg, J. Tao, J. Mendez, G. A. Rezvani, and J. J. Pekarik, "Interlaboratory Comparison of Noise-Parameter Measurements on CMOS Devices with 0.12 µm Gate Length," 66th ARFTG Microwave Measurements Conf. Digest, Washington, DC, December 2005, pp. 77-81.

J. Verspecht, D. F. Williams, D. Schreurs, K. A. Remley, and M. D. McKinley, "Linearizing Large-Signal Scattering Functions," *IEEE Trans. Microwave Theory Tech.* **53**, 1369-1376 (May 2005).

E. Gerecht, D. Gu, S. Yngvesson, F. Rodriguez-Morales, R. Zannoni, and J. Nicholson, "HEB Heterodyne Focal Plane Arrays: A Terahertz Technology for High Sensitivity Near-Range Security Imaging Systems," Proc. SPIE Defense and Security Symposium, Orlando, FL, March 2005, pp. 149-160.

A. Slavin and P. Kabos, "Approximate Theory of Microwave Generation in a Current-Driven Magnetic Nanocontact Magnetized in an Arbitrary Direction," *IEEE Trans. Magn.* **41**, 1264-1273 (April 2005).

D. F. Williams, P. D. Hale, T. S. Clement, and J. M. Morgan, "Calibrated 200 GHz Waveform Measurement," *IEEE Trans. Microwave Theory Tech.* **53**, 1384-1389 (April 2005).

F. Rodriguez-Morales, K. S. Yngvesson, E. Gerecht, N. Wadefalk, J. Nicholson, D. Gu, X. Zhao, T. Goyette, and J. Waldman, "A Terahertz Focal Plane Array Using HEB Super-Conducting Mixers and MMIC IF Amplifiers," *IEEE Micro. Wireless Comp. Lett.* **15** 199-201 (April 2005).

W. L. Johnson, S. A. Kim, S. E. Russek, and P. Kabos, "Brillouin Light Scattering From Pumped Uniform-Precession and Low-k Magnons in $Ni_{81}Fe_{19}$," *Appl. Phys. Lett.* **86**, 102507 (March 2005).

J. R. Baker-Jarvis, M. D. Janezic, B. F. Riddle, R. T. Johnk, C. L. Holloway, R. G. Geyer, and C. A. Grosvenor, "Measuring the Permittivity and Permeability of Lossy Materials: Solids, Liquids, Metals, and Negative-Index Materials," NIST Technical Note 1536 (February 2005).

T. M. Wallis, J. Moreland, B. Riddle, and P. Kabos, "Microwave Power Imaging with Ferromagnetic Calorimeter Probes on Bimaterial Cantilevers," *J. Magn. Magn. Mater.* **286**, 320-323 (February 2005).

K. A. Remley, D. F. Williams, D. Schreurs, and J. Wood, "Simplifying and Interpreting Two-Tone Measurements," *IEEE Trans. Microwave Theory Tech.* **52**, 2576-2584 (November 2004).

K. A. Remley, D. Schreurs, D. F. Williams, and J. Wood, "Extended NVNA Bandwidth for Long-Term Memory Measurements," Proc. IEEE MTT-S Int. Microwave Symp., pp. 1739-1742 (June 2004).

E. Gerecht, D. Gu, X. Zhao, J. Nicholson, F. Rodriguez-Morales, and S. Yngvesson, "Development of NbN THz HEB Mixer Coupled Through SlotRing Antennas," Proc. 15th Symp. Space Terahertz Technology, Northampton, MA, April, 2004.

P. Kabos, U. Arz, and D. F. Williams, "Multiport Investigation of the Coupling of High Impedance Probes," *IEEE Microwave Wireless Comp. Lett.* **14**, 510-512 (2004).

J. Baker-Jarvis, P. Kabos, and C. L. Holloway, "Nonequilibrium Electromagnetics: Local and Macroscopic Fields and Constitutive Relationships," *Phys. Rev. E* **70**, 036615 (2004).

J. Randa and D. K. Walker, "Amplifier Noise-Parameter Measurement Checks and Verification," 63rd ARFTG Conf. Digest, Ft. Worth, TX, pp. 41-45 (2004).

Electromagnetic Properties of Materials

Technical Contact:
Jim Baker-Jarvis

Staff-Years (FY 2006):
3.8 professionals
0.2 technician

Goals

The primary objectives of this program are to develop, improve, and analyze relevant measurement methods, uncertainties, and the underlying physics of the complex permittivity and permeability of bulk to nanoscale, naturally occurring and artificial materials in the radio-frequency through terahertz-wave spectrum as functions of temperature, frequency, and bias fields. The emphasis is on the metrology of substrates, thin films, liquids, biological materials, artificial materials, all at high frequencies and small scales.

Customer Needs

As microprocessor speeds steadily increase and new high-frequency technologies come on-line there is a continual and essential need for high-frequency materials characterization. Industries and researchers require new measurement methods on a broad array of materials with well characterized uncertainties, at microwave frequencies through terahertz, and over a range of temperatures. In addition to applications involving traditional bulk materials, trends in microwave materials are moving toward thinner layered substrates, biological materials, artificial dielectrics, and nanoscale materials.

At present, the frequencies of primary interest to the microelectronic industry are peaked in the 5 to 10 gigahertz region, but satellite communications, radars, and homeland security applications now use much higher frequencies, up into the terahertz bands. The primary driver for dielectric measurements in the microelectronics arena originates in the fact that as the operational speeds of devices increase, the dielectric losses in substrate materials severely influence microcircuit operation through signal degradation and heating.

Microelectronic circuitry is packaged on multi-layered substrates and thin films. Electronic substrate/thin film materials are used in printed wiring boards (**PWB**), low-temperature cofired ceramics (**LTCC**), central processing unit (**CPU**) chips, and microwave components. The dielectric parameters influence the propagation speed, impedance, heating, and phasing characteristics of the substrates. In order to achieve circuit miniaturization, new substrates are being developed that incorporate artificial, tunable, and layered structures. Emerging substrates that exhibit electrical properties that do not occur in naturally occurring materials are being designed with metamaterials and nanoscale composites. Such materials can provide novel device concepts useful to the commercial, military, and metrological communities. This field is in serious need of metrology, and each of these materials requires new measurement methods and well characterized uncertainties. Newly developed thin film materials such as high-temperature superconductors, ferroelectrics, and magnetoelectrics hold great potential for improved functionality in microwave devices, but are still in the critical stage of materials development. Accurate characterization at this stage of the microwave properties of these emerging materials can have a large impact on the development of future electronic systems.

Security needs and biological research require high-frequency characterization of liquids. Hence, the need for reference liquids and basic metrology has increased. Both solid and liquid dielectric reference materials are needed to provide measurement traceability to NIST. Measurement comparison provides assessments of the quality of material characterization.

Data on temperature-dependent dielectric and loss properties of ceramics, substrates, and crystals from cryogenic temperatures to 300 degrees Celsius, at microwave and millimeter frequencies, are crucial in the wireless and the time-standards arena. For example, computer-based design methods require very accurate data on the dielectric and magnetic properties of these materials over wide ranges of frequency and temperature. When interpreting measurement results, an understanding of loss mechanisms in low-loss crystals is important. Meshing of an understanding of the underlying physics with dielectric spectroscopy is becoming increasingly important in novel material research. In addition, nondestructive methods for permittivity measurement are needed throughout industry.

Various applications require composite dielectrics that simulate the human body's electrical properties for security applications such as in metal detectors and also for analyzing the effect of electromagnetic interference (**EMI**) on implanted medical devices and cell phones.

To support future needs in the microelectronics industry for the development of novel new technologies, methods for characterizing nanoscale composites will be necessary. Needs for on-chip microscale-to-nanoscale measurements of permittivity and permeability have been highlighted by the microelectronics industry.

TECHNICAL STRATEGY

The program's main thrusts are in the areas that support homeland security, biotechnology, microelectronics industry, and nanoscale metrology. To support the microelectronics industries in their quest for lower-loss materials, we will develop higher frequency measurement methods and broaden our measurement temperature range with added humidity control. Current research materials under study are liquids, thin films and printed wiring boards, low-loss dielectrics, magnetic crystals, and synthetically polarized films.

To support the nanotechnology and artificial material efforts, we will develop metrology for nanoscale composites and advanced materials such as metamaterials, multiferroics, thin films, superconductors, and ferroelectrics. We have an on-going effort focused on determining the permittivity and permeability of thin-film samples that are difficult to measure accurately with conventional cavity-based microwave measurement techniques. Material systems of interest are ferroelectric thin films and multiferroic materials, which display different properties depending on the application of an electric- or magnetic-field bias. Also of interest are thin-film materials that display relaxation behaviors within the broad frequency range of our measurement systems.

In order to understand the underlying physics of our novel material measurements, we plan to have an ongoing theoretical modeling effort that supports the interpretation of our broadband spectroscopy measurements.

In response to homeland security needs at airports, we will perform research on the identification of liquids from relaxation spectrum measurements. This research is presently funded through the NIST Office of Law Enforcement Standards.

In response to a need from the microelectronics industry for on-chip characterization, we will extend our evanescent probe system to incorporate a dielectric resonator, enhance our theoretical model, and perform measurements on relevant materials.

We will also aid the PWB and LTCC industries in measuring the permittivity of substrates at high frequencies. To this end, we will further enhance our wideband, variable-temperature metrology and extend the capability of our Fabry-Perot measurement system to include variable temperatures.

To aid the semiconductor industry for materials characterization we will measure a wide spectrum of semiconductor materials commonly used in the electronics industry as a function of temperature and frequency.

To satisfy documented needs in the health care, biotechnology, and metal-detector industries, we will characterize materials that simulate the electrical properties of the human body. In addition, in support of the biotechnology industry, we will improve our liquid measurement metrology and will compare our measurements with those of the U.K.'s National Physical Laboratory, using the liquid measurement methods we have developed.

To enhance the understanding of the physics of high-frequency losses in dielectrics, we will test ferroelectrics and crystals over wide temperature and frequency ranges using an in-house model for determination of the permittivity. We will also compare the measured losses as functions of temperature and frequency to expressions in the solid-state literature.

We will support the development of high-frequency standards by attending and contributing to IPC and IEEE standards committee meetings.

In order to investigate the possible applications of nonlinear permittivity effects we will study nonlinear material responses. Nonlinear materials are finding increased use in electronic applications. Examples include the use of ferroelectric materials as voltage-controlled components and high-temperature superconductors in high-order, low-loss filter applications. Such nonlinear materials invariably give rise to effects such as harmonic generation and intermodulation distortion, which can have important consequences at the system level. We will address these problems by evaluating the nonlinear response of broadband device structures, in order to determine the intrinsic nonlinearity of materials such as ferroelectrics and high temperature superconductors.

Evanescent microwave probe for on-chip dielectric measurements.

Accomplishments

■ **Complex Permittivities** — In response to homeland security needs and as a result of an industrial collaboration, we developed improved metrology and software for measuring the broadband complex permittivity of foams and liquids. The new software is based on an in-house theoretical model for the open-circuited sample holder. We measured the permittivity of lossy foams over K-band frequencies. We completed a Standard Reference Material (**SRM**) for both relative permittivity and loss tangent.

In a basic research study we applied broadband permittivity measurement techniques to a range of thin-film material systems, including ferroelectrics as well as voltage-tunable dielectrics and artificial multiferroics. These measurements included broadband voltage-dependent measurements of $SrTiO_3$ thin films at cryogenic temperatures, as well as measurements of $Ba_xSr_{1-x}TiO_3$ thin films at room temperature. We have applied our broadband measurement technique to artificial multiferroic thin-film samples, which are multilayers of ferroelectric $PbTiO_3$ and ferromagnetic Co-Fe-O. We performed measurements of these multiferroic samples under combined electric- and magnetic-field biases.

In order to study new approaches for permittivity determination we developed our broadband nonlinear measurement system for characterizing the broadband nonlinear microwave response. We have also developed the detailed nonlinear models used to analyze the measurement results obtained with this measurement system. Application of our measurement and analysis techniques to high temperature superconducting transmission lines has demonstrated experimentally the equivalence of different manifestations of nonlinear response (third harmonic generation, intermodulation distortion) in the same device, and has provided evidence of an inductive origin for the nonlinear response in these materials. Measurements of the nonlinear response of other ferroelectric materials at room temperature have displayed a range of behaviors, including a much larger intermodulation signal in ferroelectric $BaSrTiO_3$, a smaller microwave nonlinear capacitance compared to the DC nonlinear capacitance, and the dominance of the nonlinear conductance term in ferroelectric $PbTiO_3$ samples.

■ **Properties of Advanced/Emerging Materials** — In response to internal NIST nanowire efforts we developed a new method for measuring the electrical conductivity of bulk samples of carbon nanotubes using a dielectric resonator and in-house developed software. We performed measurements and published a paper on the resonance phase in metamaterials.

■ **Evanescent Probe Metrology and Theory** — The microelectronics industry roadmap highlights the importance of on-chip probing. In response to this need we have completed a theory for characterizing the microscopic fields present while performing on-chip permittivity measurements. In order to better understand our permittivity measurements, we developed from basic statistical mechanics the microscopic and macroscopic relationships for the entropy in electromagnetic driving; this work has been published in *Physical Review E*. We also developed a polarization evolution model that relates relaxation times to the permittivity.

■ **Semiconductor Materials** — Through conversations with industry representatives and NIST personnel, we have determined that refined characterization of semiconductors is needed. We obtained a suite of commonly used semiconductors used by researchers in the semiconductor industry and the Optoelectronics Division and performed variable frequency and variable temperature measurements on the materials. In addition, we characterized the permittivity of substrates as a function of frequency and temperature for the LTCC industry.

■ **High-Frequency Measurement System** — In order to meet the needs for measurements at higher frequencies we designed and constructed a new Fabry-Perot mirror, and preliminary measurements on two reference materials were performed. To expand on these results, a confocal Fabry-Perot system has been ordered and will also operate up to 100 gigahertz.

■ **Phase Noise** — In collaboration with the Physics Laboratory, a ceramic 10 gigahertz air-filled cavity was designed and constructed for use in the X-band phase noise measurement system. Preliminary measurements have indicated an improvement of 10 decibels for close-in phase noise. Also, design work and construction of a 40 gigahertz metallic cavity was successfully completed and incorporated into a basic discriminator circuit at 40 gigahertz. Ceramic versions of this cavity are now being constructed, with expected improvement in phase noise equivalent to the results obtained with the 10 gigahertz cavity.

■ **Standards and Comparisons** — In response to documented industry needs, a new standard for the IPC High-Frequency Task Group (D-24) has been completed and related software has been transferred to industry. In order to test our calculated uncertainties, we contributed specimens for the EUROMET International Intercomparison.

■ **Measurements on Fluids** — We have further developed a switched measurement system that allows for nearly simultaneous measurements of fluid-loaded devices using a DC meter, an LCR meter, an RF network analyzer, and a microwave vector network analyzer. We designed four-terminal measurement devices to explore the role of polarization impedances in the low-frequency measurements of conducting fluids. We further developed our modeling capabilities in order to extract permittivity values from our measurements. We have applied our measurement system to an increasingly broad range of fluid systems at different temperatures, including de-ionized water (at temperatures of 20, 37, and 50 degrees Celsius), and methanol, polystyrene beads, and NaCl buffer solutions (at 20 degrees Celsius).

Recent Publications

J. C. Booth, J. Mateu, M. D. Janezic, J. Baker-Jarvis, and J. Beall, "Broadband Permittivity Measurements of Liquid and Biological Samples Using Microfluidic Channels," Proc. IEEE MTT-S Int. Microwave Symp., San Francisco, CA, June 2006, paper TH3E-4 (in press).

J. Baker-Jarvis, M. D. Janezic, D. Love, T. M. Wallis, C. L. Holloway, and P. Kabos, "Phase Velocity in Resonant Structures," *IEEE Trans. Magn.* **42**, 3344-3346 (October 2006).

J. Baker-Jarvis, M. D. Janezic, and J. Krupka, "Measurements of Coaxial Dielectric Samples Employing Both Transmission/Reflection and Resonant Techniques to Enhance Air-Gap Corrections," Proc. 16th Int. Conf. Microwaves, Radar and Wireless Commun. (MIKON), Krakow, Poland (May 2006).

J. R. Baker-Jarvis, "Time-Dependent Entropy Evolution in Microscopic and Macroscopic Electromagnetic Relaxation," *Phys. Rev. E* **72**, 066613 (December 2005).

B. Riddle and C. Nelson, "Impedance Control for Critically Coupled Cavities," Proc. 2005 Joint IEEE Int. Freq. Control Symp., Vancouver, BC, pp. 488-493 (August 2005).

R. G. Geyer, B. Riddle, J. Krupka, L. A. Boatner, "Microwave Dielectric Properties of Single-Crystal Quantum Paraelectrics $KTaO_3$ and $SrTiO_3$ at Cryogenic Temperatures", *J. Appl. Phys.* **97**, 104111 (May 2005).

J. R. Baker-Jarvis, M. D. Janezic, B. F. Riddle, R. T. Johnk, C. L. Holloway, R. G. Geyer, and C. A. Grosvenor, "Measuring the Permittivity and Permeability of Lossy Materials: Solids, Liquids, Metals, and Negative-Index Materials," NIST Technical Note 1536 (February 2005).

R. G. Geyer, J. Baker-Jarvis, and J. Krupka "Variable-Temperature Microwave Dielectric Properties of Single-Crystal Fluorides," *Ceramic Transactions* **167**, 51-55 (2005).

J. Baker-Jarvis, P. Kabos, and C. L. Holloway, "Nonequilibrium Electromagnetics: Local and Macroscopic Fields and Constitutive Relationships," *Phys Rev E* **70**, 036615 (September 2004).

M. D. Janezic, E. F. Kuester, and J. Baker-Jarvis, "Broadband Complex Permittivity Measurements of Dielectric Substrates using a Split-Cylinder Resonator," Proc. IEEE MTT-S Int. Microwave Symp., Fort Worth, TX, pp.1817-1820 (June 2004).

M. D. Janezic and J. Baker-Jarvis, "Relative Permeability Measurements for Metal-Detector Research," NIST Technical Note 1532 (2004).

M. D. Janezic, R. F. Kaiser, J. Baker-Jarvis, and G. Free, "DC Conductivity Measurements of Metals," NIST Technical Note 1531 (2004).

M. D. Janezic, "Dielectric Measurement Methods at Millimeter-Wave Frequencies," Proc. IMAPS Conf. Ceramic Interconnect Technology, Denver, CO, pp. 221-225 (2004).

J. Baker-Jarvis, M. D. Janezic, and J. Krupka, "Broadband Dielectric Measurement of Liquids", Proc. IEEE Conf. Electrical Insulation and Dielectric Phenomena (CEIDP), Boulder, CO, pp. 17-20 (October 2004).

C. A. Grosvenor, R. Johnk, D. Novotny, S. Canales, J. Baker-Jarvis, M. Janezic, J. Drewniak, M. Koledintseva, J. Zhang, and P. Rawa, "Electrical Material Property Measurements Using a Free-Field, Ultra-Wideband System," Proc. IEEE Conf. Electrical Insulation and Dielectric Phenomena (CEIDP), Boulder, CO, pp. 174-177 (October 2004).

R. G. Geyer, J. Baker-Jarvis, and J. Krupka, "Dielectric Characterization of Single-Crystal LiF, CaF_2, MgF_2, BaF_2, and SrF_2 at Microwave Frequencies," Proc. IEEE Conf. Electrical Insulation and Dielectric Phenomena (CEIDP), Boulder, CO, pp. 493-497 (October 2004).

J. Krupka, K. Derzakowski, M. D. Janezic, and J. Baker-Jarvis, "$TE_{01\delta}$ Dielectric-Resonator Technique for Precise Measurements of the Complex Permittivity of Lossy Liquids at Frequencies Below 1 GHz," Proc. Conf. Precision Electromagnetic Measurements (CPEM), London, UK, pp. 469-470 (June 2004).

ANTENNAS AND ANTENNA SYSTEMS: ANTENNA THEORY AND APPLICATIONS

Technical Contact:
Mike Francis

Staff-Years (FY 2006):
2.0 professionals
1.0 research associate

GOALS

The Antenna Theory and Applications Project develops, refines, and extends measurement techniques to meet current requirements and to anticipate future needs for accurate antenna characterization.

CUSTOMER NEEDS

Microwave antenna hardware continues to become more sophisticated. We provide state-of-the-art measurement support for antennas and antenna systems. Customer needs include:

Improved Accuracy — High-performance systems, especially those that are satellite-based, require maintenance of tighter tolerances.

Higher Frequencies — Millimeter-wave applications up to 500 gigahertz are being developed.

Low-Sidelobe Antennas — Military and commercial communications applications increasingly require sidelobe levels of 50 decibels below peak (or better), a range where measurement by standard techniques is difficult.

Complex Phased-Array Antennas — Large, often electronically steerable, phased arrays require special diagnostic tests to ensure full functionality.

In Situ **and Remote Measurements** — Many systems cannot be transported to a measurement laboratory. Robust techniques are needed for on-site testing.

Production-Line Evaluation — Techniques are required that emphasize speed and economy, possibly at the expense of the ultimate accuracy.

RFID Reliability — RF identification (**RFID**) is being used in new applications such as inventory control and electronic passports. Test methods are needed to assure reliability and security.

Evaluation of Anechoic Chambers and Compact Ranges — A number of widely used measurement systems rely on establishing a well characterized test field. Near-field methods can be used to evaluate and analyze the quality of these test fields.

TECHNICAL STRATEGY

We seek to expand our frequency coverage for antenna calibrations to meet the demands of government and industry. A probe-position correction theory has been developed. Probe-position correction has been implemented at the NIST range by using a laser tracker to dynamically acquire position data. This will help us maintain low uncertainties as we extend the frequency coverage of our calibrations.

The near-field extrapolation method, developed at NIST, is an accurate technique to characterize the on-axis gain and polarization properties of antennas. Further improvement is still possible. We plan to extend the extrapolation software to take full advantage of phase information and to analyze the conditioning of the algorithm.

A thorough uncertainty analysis for planar near-field measurements has been previously developed. A similarly comprehensive uncertainty analysis is needed for spherical near-field measurements. We have completed a preliminary analysis and are working to refine the bounds.

In-situ near-field measurements of antenna systems are problematic because of the mechanical difficulties in maintaining position tolerances and because full spherical scans are typically not physically feasible. We have made significant progress in overcoming both these challenges and are working on the practical implementation of a deployable measurement system.

ACCOMPLISHMENTS

- **Proximity Effects in Microwave Radiometer Calibrations** — Microwave radiometers are calibrated by observing sources of known brightness temperature, assuming the source is in the far field. However, such sources are often actually in the near field. A simulation for estimating the uncertainty due to the source being in the near field of the radiometer has been completed and documented.

- **Comparison of Gains Determined from the Extrapolation and Pattern Integration Methods** — The gains of three antennas (one a dual-port probe) were determined with both the extrapolation and pattern integration methods and the results compared. For antennas with gains greater than about 15 decibels, the uncertainties of the two methods were comparable. For gains less than 15 decibels, the extrapolation method was found to be superior in the NIST facility.

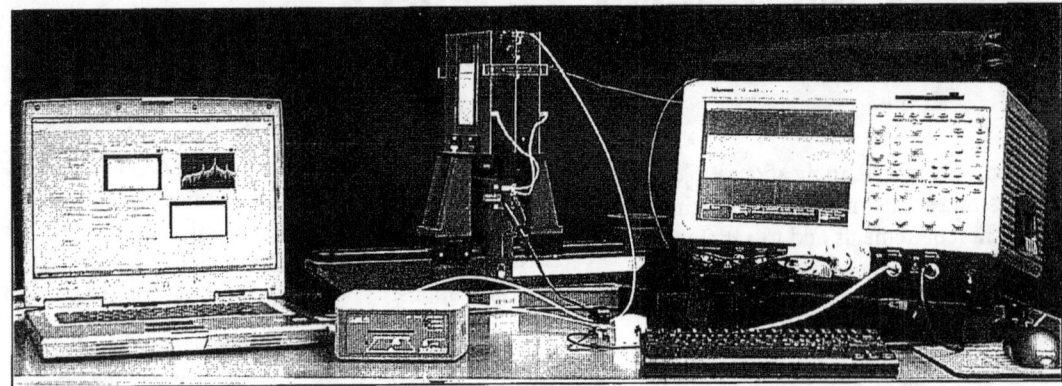

Setup for the 13.56 megahertz load modulation test, which is used to test the strength of the return signal from an RFID card.

- **RFID Test Bed** — An RFID test bed has been established at 13.56 megahertz for testing RFID cards for load modulation (return signal strength) and electromagnetic durability when the cards are exposed to electric discharge or to high strength alternating electric and magnetic fields.

Short Courses

NIST and the Georgia Institute of Technology annually offer an introductory course on antenna measurements. Every other year NIST presents an in-depth technical course restricted to near-field methods that were pioneered at NIST.

Software

Planar, cylindrical, and spherical near-field scanning applications algorithms are currently available. Probe position-correction software is available for the planar and spherical methods. The constrained least squares algorithm for partial sphere data is also available. Quiet-zone evaluation and imaging programs will be available soon.

External Recognition

- Mike Francis chairs the Antenna Standards Committee of the IEEE Antennas and Propagation Society.
- Mike Francis has been appointed to the Antenna Centre of Excellence (European Union) Scientific Council for 2006 and 2007.
- Mike Francis received the Antenna Measurement Techniques Association Distinguished Service Award in October 2004.

Recent Publications

J. R. Guerrieri, M. H. Francis, P. F. Wilson, A. B. Kos, L. E. Miller, N. P. Bryner, D. W. Stroup, and L. Klein-Berndt, "RFID-Assisted Indoor Localization and Communication for First Responders," Proc. Antenna Measurement Techniques Association (AMTA), Austin, TX (October 2006); Proc. European Conference on Antennas and Propagation, Nice, France (November 2006).

L. E. Miller, P. F. Wilson, N. P. Bryner, M. H. Francis, J. R. Guerrieri, D. W. Stroup, and L. Klein-Berndt, "RFID-Assisted Indoor Localization and Communication for First Responders," Proc. 2006 Intl. Symp. Advanced Radio Tech. (ISART), Boulder, CO, pp. 83-91 (March 2006).

M. H. Francis, K. MacReynolds, and J. R. Guerrieri, "Comparison of Gains Determined from the Extrapolation and Far-Field Pattern Integration Methods," Proc. Antenna Meas. Tech. Assoc. (AMTA), Newport, RI, pp. 304-307 (November 2005).

R. C. Wittmann, A. E. Cox, and R. H. Direen, "Proximity Effects in the Calibration of Microwave Radiometers," Proc. Antenna Meas. Tech. Assoc. (AMTA), Newport, RI, pp. 333-336 (November 2005).

J. Guerrieri, K. MacReynolds, M. Francis, R. Wittmann, and D. Tamura, "Practical Implementation of Probe-Position Correction in Planar Near-Field Measurements," Proc. Antenna Meas. Tech. Assoc. (AMTA), Atlanta, GA, pp. 356-359 (October 2004).

M. Francis, J. Guerrieri, K. MacReynolds, and R. Wittmann, "Estimating Multiple-Reflection Uncertainties in Spherical Near-Field Measurements," Proc. Antenna Meas. Tech. Assoc. (AMTA), Atlanta, GA, pp. 85-87 (October 2004).

R. Wittmann and M. Francis, "Spherical Scanning Measurements: Propagating Errors Through the Near- to Far-Field Transform," Proc. Antenna Meas. Tech. Assoc. (AMTA), Atlanta, GA, 74-79 (October 2004).

R. Wittmann, B. Alpert, and M. Francis, "Near-Field Spherical-Scanning Antenna Measurements with Nonideal Probe Locations," *IEEE Trans. Ant. Prop.* **52**, 2184 (August 2004).

Antennas and Antenna Systems: Antenna Near-Field Measurements

Technical Contact:
Katie MacReynolds

Staff-Years (FY 2006):
2.0 professionals
1.0 technician

Goals

The Antenna Near-Field Measurements Project serves as a national resource by providing antenna measurement services and traceability through calibrations. It supports government and private industry programs, and maintains and develops the standards, methods, and instrumentation for antenna characterization of gain, polarization and pattern measurements.

Customer Needs

We continue to upgrade antenna metrology capability to meet evolving customer demands in the following areas:

Probe Characterization — Accurate probe characterization is fundamental to precise antenna measurements. We provide probe correction coefficients for use in planar, spherical and cylindrical near-field facilities.

Planar and Spherical Near-Field Measurements — These are required to accurately characterize large aperture, high-frequency antennas such as phased array and dish antennas used in satellite communications.

Antenna Standard Characterization — Industry and government require antenna standards for in-house antenna measurements.

Measurement Traceability — Program specifications often require NIST traceability.

Independent Verification of Antenna Parameters — Government and industry request measurements to verify that their measurement and analysis procedures produce the predicted and correct results.

Technical Support — Assistance on measurement techniques and analysis algorithms for antenna facilities that are implementing near-field measurements.

Technical Strategy

We currently maintain near-field antenna measurement standards and capabilities for frequencies from 1.5 to 110 gigahertz. The recent integration of a laser tracker system provides accurate information on probe position for use with position-correction algorithms. These improvements will help maintain low uncertainties as measurement frequencies are increased in the future.

We have extended the frequency range of our extrapolation measurement capability. The new extrapolation range provides on-axis gain characterization capability to complement near-field antenna pattern measurements at millimeter-wave frequencies. Plans to provide polarization characterization up to 110 gigahertz are in process and will be available in the near future.

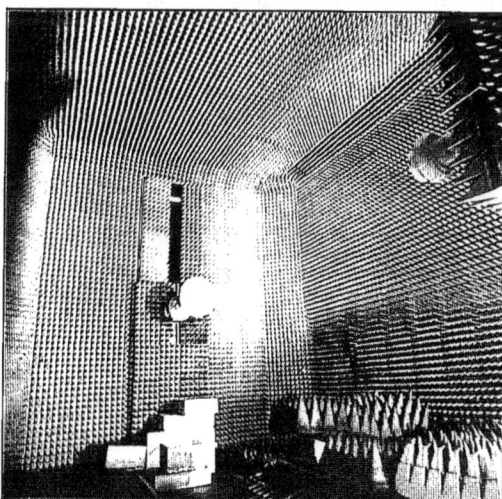

The NIST probe pattern range with the fixed probe located near the center of the photo and the moving probe located on the moving tower in the upper right.

We are performing an internal comparison on all three of the NIST near-field ranges — planar, cylindrical, and spherical — to verify performance and refine uncertainties. Our capability with all three types of ranges is unique.

Accomplishments

■ **Antenna Calibrations** — We perform antenna calibrations for external customers. Examples of recent measurements are:

- WR-28 — open-ended waveguide probe
- WR-62 — 3 standard gain horns
- WR-90 — 3 standard gain horns and an open-ended waveguide probe
- WR-42 — 2 standard gain horns and 2 dual-port circularly polarized probes

- WR-22 — 2 standard gain horns and 2 dual-port circularly polarized probes
- WR-62 — Cassegrain dish antenna

■ A 60 to 110 gigahertz extrapolation range has been designed and installed. Antenna gain tests at 94 gigahertz show excellent performance. The new range significantly extends NIST antenna metrology capability.

Recent Publications

J. R. Guerrieri, M. H. Francis, P. F. Wilson, A. B. Kos, L. E. Miller, N. P. Bryner, D. W. Stroup, and L. Klein-Berndt, "RFID-Assisted Indoor Localization and Communication for First Responders," Proc. Antenna Measurement Techniques Association (AMTA), Austin, TX (October 2006); Proc. European Conference on Antennas and Propagation, Nice, France (November 2006).

K. MacReynolds, J. Guerrieri, and D. Tamura, "Three-Antenna Extrapolation Gain Measurement System at Millimeter Wave Frequencies," Proc. Antenna Meas. Tech. Assoc. (AMTA), Austin, TX (October 2006).

L. E. Miller, P. F. Wilson, N. P. Bryner, M. H. Francis, J. R. Guerrieri, D. W. Stroup, and L. Klein-Berndt, "RFID-Assisted Indoor Localization and Communication for First Responders," Proc. 2006 Intl. Symp. Advanced Radio Tech. (ISART), Boulder, CO, pp. 83-91 (March 2006).

J. Guerrieri, K. MacReynolds, M. Francis, R. Wittmann, and D. Tamura, "Planar Near-Field Measurement Results at 94 GHz Using Probe Position Correction," Proc. Antenna Meas. Tech. Assoc. (AMTA), Atlanta, GA, pp. 356-359 (October 2005).

M. Francis, K. MacReynolds, and J. Guerrieri, "Comparison of Gains Determined from the Extrapolation and Pattern Integration Methods," Proc. Antenna Meas. Tech. Assoc. (AMTA), Atlanta, GA, pp. 356-359 (October 2005).

J. Guerrieri, K. MacReynolds, M. Francis, R. Wittmann, and D. Tamura, "Practical Implementation of Probe-Position Correction in Planar Near-Field Measurements," Proc. Antenna Meas. Tech. Assoc. (AMTA), Atlanta, GA, pp. 356-359 (October 2004).

J. Randa, A. Cox, D. Walker, M. Francis, J. Guerrieri, and K. MacReynolds, "Standard Radiometers and Targets for Microwave Remote Sensing," Int. Geoscience and Remote Sensing Symp. (IGARSS), Anchorage, AK (September 2004).

J. Guerrieri, K. MacReynolds, and D. Tamura, "Evaluation of a Radiometer Phase Retardation Plate Using Planar Near-Field Measurements," Proc. National URSI Meeting, Boulder, CO (January 2004).

J. Guerrieri, K. MacReynolds, and D. Tamura, "Radiometer Phase Retardation Plate Evaluation Using Planar Near-Field Antenna Measurements," Proc. Antenna Meas. Tech. Assoc. (AMTA), Irvine, CA (October 2003).

Antennas and Antenna Systems: Metrology for Radar Cross Section Systems

Technical Contact:
Lorant Muth

Staff-Years (FY 2006):
1.0 professional

Goals

The Metrology for Radar Cross Section Systems Project assists the U.S. Department of Defense (**DOD**) and industrial radar cross section (**RCS**) measurement ranges to create and implement a National DOD Quality Assurance Program to ensure high-quality RCS calibrations and measurements with stated uncertainties.

Customer Needs

RCS measurements on complex targets, such as aircraft, ships, and missiles, are made at different types of RCS measurement ranges, including compact ranges (indoor static), and outdoor static or dynamic facilities. Measurements taken at various ranges on the same targets must agree with each other within stated uncertainties to increase confidence in RCS measurements industry-wide. Although the sources of uncertainty are well known, a comprehensive determination of the magnitudes of uncertainties in RCS calibrations require well formulated procedures that measurement ranges can use to determine their uncertainties. Customer needs include:

Calibration Artifacts — RCS users need improved calibration artifacts that are dimensionally traceable and calculable and exhibit wide dynamic range.

Calibration Procedures — Calibration procedures and data analysis techniques are needed to minimize range uncertainties at both government and industrial RCS ranges. The implementation of improved procedures and the determination of range uncertainties at every RCS measurement range are essential if the U.S. RCS industry is to maintain its world leadership.

Technical Strategy

The complex measurement systems and measurement practices at RCS ranges should be documented uniformly throughout the industry so that meaningful comparison of capabilities and important range-to-range differences are recognized. The framework of a RCS Range Book, in the context of a DOD RCS Self-Certification Program, is used to ensure community-wide compliance.

We provide RCS Range Book reviews for the DOD and industrial RCS ranges. These in-depth reviews provide guidance to the RCS community as they pursue their industry-wide certification program. The uncertainty analyses pursued by the U.S. RCS ranges are based on the pioneering NIST work in this area.

We have continued to work closely with selected RCS measurement ranges to develop and standardize procedures to determine RCS calibration and measurement uncertainty for both monostatic and bistatic RCS measurements. Fully polarimetric calibration procedures are also being studied.

To support these research activities we have recommended an expanded set of RCS calibration cylinders to calibrate the system at various signal levels of interest using a single artifact. To support polarimetric calibration research, we recommended a set of calibration dihedrals that can be used to determine system parameters needed to analyze polarimetric calibration data.

We seek to fully assess the technical merit and deficiencies of existing calibration and measurement procedures, data-analysis techniques, and uncertainty analysis. We plan to publish recommendations for improvements in these areas. We plan to further explore known problems in areas such as dynamic sphere calibration, polarimetric calibration, and bistatic RCS calibration.

The annual RCS Certification Meeting held at NIST-Boulder provides a forum for the RCS community to discuss procedural and technical issues on an ongoing basis.

Accomplishments

- **Cylinder Calibration Set** — The RCS community has adopted a basic cylinder calibration set to test the calibration integrity of monostatic RCS systems. Computed radar cross sections for the cylinder set have been obtained. These four cylinders have been measured at a number of government and industrial measurement ranges, with agreement with the theoretical RCS of better than 0.5 decibels. These comparisons demonstrate good repeatability; however, we need more robust independent measurement procedures to determine the measurement uncertainties.

- **Dihedral Calibration Set** — We designed and manufactured a set of calibration cylinders with dihedral cutouts that can be used to calibrate an RCS range within a large dynamic range rather than at a single signal level, thereby improving calibration accuracy within the measurement interval. These artifacts should be useful in a nationwide measurement-comparison program.

Sample set of calibration dihedrals.

■ **RCS Uncertainty Analysis** — We have completed a measurement-based RCS calibration uncertainty analysis for the Etcheron Valley Range, NAVAIR, China Lake, CA. This study determined the calibration uncertainty bounds without having to rely on statistical model assumptions that may not be valid for RCS calibrations and measurements.

■ **Polarimetric Calibrations** — RCS ranges have reported less-than-satisfactory results with existing polarimetric calibration procedures. We developed a more robust calibration procedure wherein full polarimetric data are obtained using a dihedral rotating around the line-of-sight to the radar. The new procedure allows us to: (1) improve the signal-to-noise ratio and check for alignment problems by exploiting the symmetry properties of the dihedral, (2) correct for the overall angular bias in the rotation angle, and (3) remove the effects of drift to obtain drift-free system parameters.

Recent Publications

L. Muth "Cross-Polarization Parameters in the Presence of Drift in Radar Cross Section Measurements," Proc. Antenna Meas. Tech. Assoc. (AMTA), Austin, TX, (October 2006) (in press).

L. A. Muth, C. M. Wang, and T. Conn, "Robust Separation of Background and Target Signals in Radar Cross Section Measurements," *IEEE Trans. Instrum. Measure.* **54**, 2462-2468 (December 2005).

L. Muth, C. Williams, D. Morales, and T. Conn, "Angular Errors in Polarimetric Radar Cross Section Calibration Using a Rotating Dihedral," Proc. Antenna Measurement Techniques Assoc. (AMTA), Providence, RI (October 2005).

L. Muth, C. Johnson, D. Morales, and T. Conn, "Angular Errors in Polarimetric Radar Cross Section Calibration Using a Rotating Dihedral," NIST Technical Note 1539 (April 2005).

L. Muth, D. Diamond, and J. A. Lelis, "Uncertainty Analysis of Radar Cross Section Calibrations at Etcheron Valley Range," NIST Technical Note 1534 (September 2004).

ELECTROMAGNETIC COMPATIBILITY: REFERENCE FIELDS AND PROBES

Technical Contact:
Perry Wilson

Staff-Years (FY 2006):
3.0 professionals

GOALS

The Reference Fields and Probes Project develops methods and techniques for establishing continuous-wave electromagnetic (**EM**) reference fields and transfer probes for frequencies to 100 gigahertz. It maintains the capability to provide antenna, probe, and field measurements with international comparability and traceability to NIST in support of U.S. industry. Although most present applications utilize spectra in the 1 megahertz to 10 gigahertz range, systems such as automotive collision avoidance radars that operate up to nearly 100 gigahertz are being developed.

CUSTOMER NEEDS

Based on the principles of "one product, one technically valid international standard, one conformity assessment" (1998 MSL Strategic Plan), industry requires EM field measurement capabilities and transfer probes that are traceable to NIST in order to meet multinational compliance requirements and reduce barriers to worldwide acceptance of U.S. products. We address these needs with the following:

Reference Fields — Well defined EM reference fields are necessary for the calibration of antennas and probes. They are also needed for research and development to increase measurement accuracy and spectral range as will be necessary to support the future needs of U.S. industry and private test laboratories.

Field Probes — Accurate field probes are needed by government and industry to define EM field levels. U.S. defense and homeland security agencies rely heavily on EM systems for sensors and strategic communication. New probes need to be developed for the ever-expanding range of EM environments.

Probe Calibrations — Field probe calibrations are costly. Techniques to reduce calibration costs are needed, especially for applications that require multiple probes and frequent recalibration.

TECHNICAL STRATEGY

We maintain an integrated effort both to generate standard reference fields and to develop the probes required for their accurate measurement. The two efforts complement each other and allow cross checking in order to reduce the uncertainties inherent in each effort as well as to transfer calibration capabilities to other test laboratories and facilities. As instrumentation and electronics achieve higher clock rates, measurements are needed at higher frequencies. We are working both to extend current techniques and facilities to higher frequencies and to develop new test methods to increase accuracy and reduce measurement costs. In this context, we plan to develop improved methods for measuring radio frequency (**RF**) emissions above 1 gigahertz.

Open area test site (**OATS**) facilities are accepted as standard sites for electromagnetic compatibility (**EMC**) emissions measurements. We are working on improvements of antenna characterization through tighter standards documentation, updated technology and enhanced methodology for EMC antenna measurements. We work closely with the American National Standards Institute (**ANSI**) and the Society of Automotive Engineers (**SAE**) to further improve their methods for EMC antenna measurements.

Fully anechoic chamber (**FAC**) facilities are accepted as standard sites for free-space measurements. Time-domain techniques are being studied as a way to measure the characteristics of these rooms and to improve the results obtained within. This type of chamber is also being evaluated for EMC product testing up to 40 gigahertz.

Closed test systems such as transverse electromagnetic (**TEM**) cells have been widely adopted for testing small antennas, sensors, and probes, but are normally limited by geometrical constraints to frequencies below 1 gigahertz. We are currently constructing a new closed-cell system that utilizes a co-conical geometry that can be used to test such devices up to 45 gigahertz. The test volume of this system is large enough to calibrate several probes at once. EM modeling and analysis using numerical techniques such as finite-difference time-domain are used to predict system performance for multiple probes.

Measurements performed by use of different equipment and facilities such as OATS, TEM cells, FAC and semi-anechoic facilities often yield different results. We will focus on systematically investigating methods to reduce these variations and improve agreement within the U.S. industrial community.

We will provide technical information and guidance to standards organizations to help correlate measurements between various EMC test facilities. We will also cooperate with the national test laboratories of our international trading partners to perform round-robin testing and comparison of various standard antennas and probes. This assures international agreement in their performance and reduces the uncertainties in the areas of metrology that affect international trade.

Accomplishments

- **Probe Calibrations** — Calibrations were performed on probes/antennas for several companies and/or government agencies covering the frequency range of 10 kilohertz to 45.5 gigahertz by use of TEM cell and anechoic chamber test facilities. Field levels varied from 1 to 200 volts per meter.

- **Co-Conical Field Generation System (CFGS)** — A 10 megahertz to 40 gigahertz RF probe test facility for the U.S. Air Force is nearing completion. The final machining of the test cell was completed in July 2006 and delivered to NIST for testing and full system integration. The CFGS is made of two major subsystems. The first is a harmonically pure RF generating system capable of producing 25 watts of RF power out of a single RF connector from 10 megahertz to 40 gigahertz. This subsystem involved the design of new RF components that are now standard product offerings from two companies. The power delivery subsystem feeds the power into the second subsystem: a broadband TEM transmission system and termination that can generate high-intensity fields and calibrate probes faster, and with comparable uncertainties, than conventional anechoic chambers.

Co-conical field generation system being assembled at NIST in preparation for delivery to the U.S. Air Force. The height of the cone is 3 meters.

- **Electric and Magnetic Field Probe** — A loop antenna with integrated photonics and controls to simultaneously measure electric and magnetic fields at levels up to 1 kilovolt per meter has been developed.

Recent Publications

C. L. Holloway, P. F. Wilson, and R. F. German, "The OATS Method Revisited," *IEEE EMC Newsletter*, pp. 41-43 (August 2005).

J. Hamilton and K. Masterson, "Accurate Bias Point Control for an Electrically Isolated Mach-Zehnder Interferometric Modulator Via an Analog Optical-Fiber Link," Proc. SPIE, Denver, CO, **5531**, 323-331 (August 2004).

P. Wilson, C. Holloway, and G. Koepke, "A Review of Dipole Models for Correlating Emission Measurements Made at Various EMC Test Facilities," Proc. 2004 IEEE Int. Symp. Electromagnetic Compatibility, Santa Clara, CA, pp. 898-901 (August 2004).

Electromagnetic Compatibility: Complex Fields

Technical Contact:
Galen Koepke

Staff-Years (FY 2006):
3.0 professionals
2.2 research associates

Goals

The Complex Fields Project develops and maintains measurement methods to quantify fields in complex environments, such as electrically large cavities and highly nonuniform boundaries. Applications include the reverberation chamber, the statistics of electromagnetic fields in rooms and buildings, the communications needs of first responders to emergencies, measurements of shielding effectiveness of advanced composites, coupling to large-scale systems and components, coupling to biological objects, advanced numerical methods, and metamaterials. These efforts support industry and government agencies, national and international standards, health care, homeland security, and nanotechnology.

Customer Needs

Large, complex systems located in complex field environments need to be tested for electromagnetic compatibility (**EMC**). Electromagnetic interference (**EMI**) affects U.S. competitiveness (through trade restrictions and regulations), national security, health, and safety. EMC regulations and requirements constitute 1 to 10 percent of the total U.S. product costs and can cause delays to market. We are providing research and support to address a number of critical areas:

Reverberation Chambers — Reverberation chambers are used to test large complex systems at high frequencies. Users of reverberation chambers need better models of statistical parameters, improved chamber and stirrer designs, test object models, and guidance for incorporating reverberation chamber technology into national and international standards.

Shielding of Advanced Composites — Shielding effectiveness of advanced composites cannot be readily tested with existing methods, such as the American Society for Testing and Materials (**ASTM**) coaxial fixture. New methods are needed, such as the use of nested reverberation chambers.

Coupling to Biological Objects — The increased use of wireless devices in scattering-rich environments is creating questions about the effects of such electromagnetic field exposure on humans and animals. These conditions need to be replicated in controlled experiments.

First Responder Communications — First responders to emergency situations encounter difficult communications scenarios. There is a need to better understand propagation in a wide variety of environments, including collapsed buildings.

Complex Boundaries — The electromagnetic interface between complex media is a difficult modeling and measurement problem. Advances at both the macro- and micro-scale are needed.

Technical Strategy

Our goal is to develop and evaluate reliable and cost-effective standards, test methods, and measurement services related to complex electromagnetic fields for EMC of electronic devices and other applications in health, defense, and homeland security. This includes investigating new applications for existing test facilities as well as improving methods for evaluating the critical characteristics of support hardware, such as antennas, cables, connectors, enclosures, and absorbing material.

Reverberation chambers are increasingly a key tool for EMC testing in the gigahertz frequency range. The recent publication of IEC61000-4-21 on reverberation chamber test methods will increase usage. We have played a leading role in developing reverberation chamber technology. We continue to develop models for the statistical behavior of the fields in the test volume and near the boundaries. While the typical target for a reverberation chamber is a Rayleigh field distribution, multiple-input/multiple-output (**MIMO**) system testing requires a Rician field distribution. We will investigate methods to accurately control the ratio between direct and indirect coupling for MIMO and other test applications. Probes are traditionally calibrated in highly controlled reference fields. We will investigate whether the large volume of a reverberation chamber can be used to simultaneously calibrate a large number of field probes. We are developing analytical models of test-object directivity. We will continue to experimentally test these models and transfer results to committees developing standards for reverberation chambers.

Advanced composites offer weight and performance advantages over metals and are increasingly being used in aerospace and other applications. Plastics inherently provide no significant shielding to electromagnetic fields. Plastics can be "metalized" by introducing conducting fibers; however, this may affect mechanical performance. There is a need to reliably measure the EM shielding properties of advanced composites so that manufacturers can find the right balance between electrical and mechanical performance for a particular application. We are investigating the use of nested reverberation chambers for this purpose. We are investigating better statistical descriptors for the shielded fields to more accurately define shielding effectiveness in complex coupling environments.

With the proliferation of wireless devices in recent years, there is a growing need to test the operation and functionality of these various devices in different multipath environments, ranging from line-of-sight environment to a pure Rayleigh environment. We have recently shown how a reverberation chamber can be used to generate a variable K-factor propagation environment for the testing of wireless communications devices. It was shown that by judiciously changing the characteristics of the reverberation chamber and/or the antenna configurations in the chamber, any desired Rician K-factor (that is, any desirable Rician multipath propagation environment) could be achieved. We have developed two different approaches, the first based on one transmitting antenna, and the second based on two transmitting antennas. Reverberation chambers represent reliable and repeatable test facilities that have the capability of simulating any Rician multipath environment for the testing of wireless communications devices. Such a test facility will be useful in the testing of the operation and functionality of the new emerging wireless devices in the future. Our paper on this topic was published in November 2006 issue of *IEEE Transactions on Antennas and Propagation*. In it, we laid out the framework, presented a few simple expressions, and presented experimental evidence to support the possibility of using reverberation chambers for testing of wireless devices in different Rician multipath environments. Not all multipath environments can be characterized by a Rician distribution. An example would be an environment where both scattering components and multiple direct components are present. In such an environment, the direct coupling components have different amplitudes and phases (or different Doppler shifts). Such environments could in principle be simulated in reverberation chambers in which several transmitting antennas are used. The use of reverberation chambers in these other multipath environments is the topic of future work.

First responders need reliable communications in emergency scenarios. Disaster scenarios and terrorist attacks may result in scenarios where responders or citizens are trapped in collapsed or blocked buildings. The propagation of signals in the bands used by first-responder radios and cellular telephones needs to be investigated. We have performed unique experiments to define communications links in buildings prior, during, and after demolition. These data will give invaluable insight into the communications problems faced by first responders.

Metamaterials (that is, engineered or man-made materials) have generated considerable interest in recent years. Metamaterials are commonly engineered by arranging a set of scatterers throughout a region of space in a specific pattern so as to achieve some desirable bulk behavior. Examples of electromagnetic metamaterials are artificial dielectrics, photonic bandgap structures, and frequency-selective surfaces. Recently there have been studies on the properties and potential applications of double-negative (**DNG**) materials. We are investigating a composite medium consisting of insulating magneto-dielectric spherical particles embedded in a background matrix to achieve DNG behavior. We have shown that the effective permeability and permittivity of the mixture can be simultaneously negative for wavelengths where the spherical inclusions are resonant.

We work closely with national and international standards bodies to transfer experimental and theoretical results and to improve test methods for large, complex systems. We plan to continue participation in various IEC, CISPR, ANSI, SAE and IEEE standards committees related to EMC test methods.

Division staff members map signals from transmitters buried in a collapsed building.

Electromagnetics Division

Accomplishments

- **Phantom Study** — The electric field statistics in a heavily loaded reverberation chamber were evaluated and compared to the empty-chamber statistics. The load consisted of phantoms (water bottles) filled with tissue-simulating liquid. Whole-body average specific absorption rate (**SAR**) in the phantoms was also determined via direct measurements of temperature increase and indirect measurements of insertion loss. The results support the use of a reverberation chamber for animal exposure studies.

- **Propagation in Collapsed Buildings** — Three experimental studies on signal propagation in buildings prior, during, and after demolition were completed. These experiments provide invaluable data on first-responder communication challenges.

- **Effective Boundary Conditions** — Effective boundary conditions for thin films with applications to active materials such as frequency-tunable surfaces have been derived and published.

- **Equivalent-Layer Models** — Three equivalent-layer models were developed for the electromagnetic characterization of carbon fiber composites used on aircraft.

Recent Publications

C. L. Holloway, G. Koepke, D. Camell, K. Remley, and S. Schima, "Shielding and Attenuation Properties of Large Buildings and Structures," Proc. 2006 EMC Europe Conf., Barcelona, Spain, pp. 465-469 (September 2006).

J. Coder and John Ladbury, "Cable Shielding Measurements Based on a Reference Unshielded Cable, or, Do You Really Want to Base Your Results on a Naked Cable?" Proc. 2006 IEEE Int. Symp. Electromagnetic Compatibility, Portland, OR, paper WE-AM-1-7 (August 2006).

D. Love, J. Ladbury, and E. Rothwell, "Comparing Numerical and Experimental Results for the Shielding Properties of a Doubly-Periodic Array of Apertures in a Thick Conducting Screen," Proc. 2006 IEEE Int. Symp. Electromagnetic Compatibility, Portland, OR, paper TH-PM-1-2 (August 2006).

C. L. Holloway, G. Koepke, D. Camell, K. A. Remley, S. Schima, and R. T. Johnk, "Propagation and Detection of Radio Signals Before, During and After the Implosion of a Large Convention Center," NIST Technical Note 1542 (May 2006).

C. L. Holloway, D. A. Hill, J. M. Ladbury, and G. H. Koepke, "Requirements for an Effective Reverberation Chamber: Unloaded or Loaded," *IEEE Trans. Electromagn. Compat.* **48**, 187-194 (February 2006).

C. L. Holloway, M. A. Mohamed, E. F. Kuester, and A. Dienstfrey, "Reflection and Transmission Properties of a Metafilm: With an Application to a Controllable Surface Composed of Resonant Particles," *IEEE Trans. Electromagn. Compat.* **47**, 853-865 (November 2005).

C. L. Holloway, M. S. Sarto, and M. Johansson, "Analyzing Carbon Fiber Composite Materials with Equivalent-Layer Models," *IEEE Trans. Electromagn. Compat.* **47**, 833-844 (November 2005).

C. L. Holloway, G. H. Koepke, D. G. Camell, K. A. Remley, D. Williams, S. Schima, and S. Canales, "Propagations and Detection of Radio Signals Before, During and After the Implosion of a Large Sports Stadium (Veterans' Stadium in Philadelphia)," NIST Technical Note 1541 (October 2005).

C. L. Holloway, G. H. Koepke, D. G. Camell, K. A. Remley, D. Williams, S. Schima, S. Canales, and D. T. Tamura, "Propagation and Detection of Radio Signals Before, During and After the Implosion of a 13 Story Apartment Building," NIST Technical Note 1540 (May 2005).

M. Johansson, C. L. Holloway, and E. F. Kuester, "Effective Electromagnetic Properties of Honeycomb Composites, and Hollow-Pyramidal and Alternating-Wedge Absorbers," *IEEE Trans. Ant. Propagat.*, **53**, 728-736 (February 2005).

J. Baker-Jarvis, P. Kabos, and C. L. Holloway, "Nonequilibrium Electromagnetics: Local and Macroscopic Fields and Constitutive Relationships," *Phys Rev E* **70**, 036615 (September 2004).

P. Wilson, "Test Object Electrical Size and Its Implication on Pattern Sampling," Proc. 2004 IEEE Int. Symp. Electromagnetic Compatibility, Santa Clara, CA, pp. 349-352 (August 2004).

Electromagnetic Compatibility: Time-Domain Fields

Technical Contact:
Robert Johnk

Staff-Years (FY 2006):
2.0 professionals
1.0 research associate

Goals

The Time-Domain Fields Project develops basic metrology and measurement techniques for a wide variety of applications such as antenna and sensor calibrations, evaluation of electromagnetic compatibility (**EMC**) measurement facilities, shielding performance of aircraft, nondestructive testing of electrical material properties, precise generation of standard fields, and detection of signals and threats.

Customer Needs

Time-domain field methods use time windowing to eliminate unwanted signals. These methods find application to problems that cannot readily be evaluated using traditional continuous-wave radiated tests. In particular, time-domain methods allow for localization of large systems and for testing in highly cluttered environments. Customer needs include:

Reflection Properties — EMC test sites, such as anechoic chambers and open area test sites (**OATS**) use absorbing and low-reflectivity materials to achieve desired performance. The reflectivity of these materials needs to be accurately characterized over a wide frequency range, possibly *in situ*.

Shielding Effectiveness — Time-domain signals can be used to investigate the shielding effectiveness of large, complex geometries. Aircraft can be tested in situ, either over tarmac or in a hanger.

Propagation in Buildings — Buildings present complicated propagation environments. Communications system designers need to know how waves couple through differing building materials and how waves couple from the exterior to interior and between interior locations.

Ultra-Wideband Systems — Ultra-wideband systems are being proposed to increase capacity and to address advanced communications needs. Antenna characteristics and link performance need to be accurately determined.

Technical Strategy

We have developed measurement tools and systems both to generate and receive ultra-wideband radiated signals. We currently have a large cone and ground plane facility. This facility can be used to generate well defined pulse fields for sensor calibration up to 20 gigahertz, plus antenna calibration and system characterization. We used extensive numerical simulations to optimize the feed and cone sections to achieve very high performance levels.

We have been a leader in developing transverse electromagnetic (**TEM**) horn antennas for receiving ultra-wideband signals. These antennas have very linear phase characteristics and are able to accurately preserve time-domain traces. We will publish a NIST technical note that summarizes our research in TEM Horns.

TEM horns are used in field-deployable systems for transmitting and receiving time-domain fields. The system uses optical fiber links to achieve high isolation and dynamic range. The system has been successfully applied to the evaluation of the shielding effectiveness properties of commercial and military aircraft. These efforts have assisted NASA, U.S. manufacturers, the U.S. Department of Defense (**DOD**), and the Federal Aviation Administration (**FAA**), to improve flight safety and reduce the vulnerability of aircraft to electromagnetic interference and threats.

We are applying time-domain techniques to develop a database of the electrical properties of building materials in support of homeland security goals. The database will help first responders develop communication systems that will improve performance in emergency situations. We are collaborating on the use of genetic algorithms to accurately extract electrical parameters from time-domain reflection and transmission data. We are also deploying a portable measurement system to measure the propagation characteristics of buildings over ultra-wide bandwidths and provide both time- and frequency-domain channel characteristics.

We have supported industry and standards groups in assessing the performance of electromagnetic facilities such as anechoic and semi-anechoic chambers, shielded rooms, reverberation chambers, and OATS facilities. The performance of microwave absorbers has been measured. Many facilities use low-reflectivity material for test object support and weather protection. We are helping to develop time-domain-based test methods to determine the effects of "reflectionless" materials of the type specified in many radiated test standards on site ability to perform ultra-wideband RCS measurements.

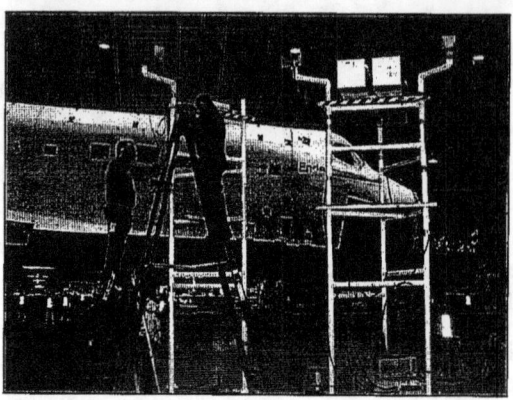

Division staff members perform field penetration tests on a NASA space orbiter.

ACCOMPLISHMENTS

■ **Aircraft Shielding** — We have developed a unique, leading-edge portable measurement system for the evaluation of coupling to complex aircraft environments. The system has been used to evaluate a number of flight vehicles such as the NASA Orbiter Endeavour, a U.S. Air Force early-warning aircraft, and a Boeing 767 commercial aircraft. Evaluations of a Boeing 737-200 and a Bombardier Global 5000 aircraft at the Federal Aviation Administration (**FAA**) Technical Center in Atlantic City, New Jersey, were recently performed. The data from this effort are in the public domain and have been disseminated to the aircraft testing community.

■ **Complex Cavity Modeling** — Numerical models of complex cavities have been developed to facilitate understanding of measurements performed in aircraft environments. This work is sponsored by the FAA.

■ **Building Materials Characterized** — A free-field materials measurement system for the evaluation of building materials has been developed.

■ **EMC Facilities Tested** — Several EMC facilities have been tested. A measurement effort was conducted recently at a wastewater instrumentation company to assess the effectiveness of an anechoic chamber retrofit. Significant improvements in chamber performance were realized through a NIST portable measurement system.

RECENT PUBLICATIONS

R. Johnk, D. Novotny, C. Grosvenor, and N. Canales, "An Electric-Field Uniformity Study of an Outdoor Vehicular Test Range," IEEE Int. Symp. Electromagnetic Compatibility, Portland, OR (August 2006).

R. T. Johnk, D. R. Novotny, C. A. Grosvenor, N. Canales, and J. G. Veneman, "Time-Domain Measurements of Radiated and Conducted Ultrawideband Emissions," *IEEE Aerospace and Electronic Systems Mag.*, **19**, 18-22 (August 2004).

D. Novotny, C. Grosvenor, R. Johnk, and N. Canales, "Panoramic, Ultrawideband, Diagnostic Imaging of Test Volumes," Proc. 2004 IEEE Int. Symp. Electromagnetic Compatibility, Santa Clara, CA, pp. 25-28 (August 2004).

C. Grosvenor, R. Johnk, D. Novotny, and N. Canales, "TEM Horn Antennas: A Promising New Technology For Compliance Testing," Proc. 2004 IEEE Int. Symp. Electromagnetic Compatibility, Santa Clara, CA, pp. 913-918 (August 2004).

Magneto-Mechanical Measurements for High Current Applications

Goals

This project specializes in measurements of the effect of mechanical strain on materials for high-current applications. Recent research has produced the first electromechanical data for the new class of high-temperature coated superconductors, one of the few new technologies expected to have an impact on the U.S. electrical power grid and the large electric power industry. The project's research has also lead to the first four patents on contacts for high-temperature superconductors.

Recent research also includes extending the high-magnetic-field limits of electromechanical measurements for development of nuclear-magnetic-resonance (**NMR**) spectrometers operating at 23.5 teslas and 1 gigahertz, and the next generation of accelerators for high-energy physics.

The Strain Scaling Law, previously developed by the project for predicting the axial-strain response of low-temperature superconductors in high magnetic fields, is now being extended to high compressive strains for use in finite-element design of magnet structures.

Project staff members in the laboratory for superconductor strain measurements.

Customer Needs

The recent success of the "second generation" of high-temperature superconductors has brought with it new measurement problems in handling these brittle conductors. We have the expertise and equipment to address these electromechanical problems. Stress and strain management is one of the key technology areas needed to move the second-generation high-temperature coated conductors to the market place. The project utilizes the expertise and unique electromechanical measurement facilities at NIST to provide performance feedback and engineering data to companies and national laboratories fabricating these conductors in order to guide their decisions at this critical scale-up phase of coated-conductor development.

The project serves industry primarily in two areas. First is the need to develop a reliable measurement capability in the severe environment of superconductor applications: low temperature, high magnetic field, and high stress. The data are being used, for example, in the design of magnets for the magnetic-resonance-imaging (**MRI**) industry, which provides invaluable medical data for health care, and contributes $2 billion per year to the U.S. economy.

The second area is to provide data and feedback to industry for the development of high-performance superconductors. This is especially exciting because of the large effort being devoted to develop superconductors for grid reliability and enhanced power-transmission capability. We receive numerous requests, from both industry and government agencies, for accurate electromechanical data to help guide their efforts in research and development in this decisive growth period.

Technical Strategy

Our project has a long history of unique measurement service in the specialized area of electromechanical metrology. Significant emphasis is placed on an integrated approach. We provide industry with first measurements of new materials in areas where there is significant research potential.

Electromechanical Measurements of Superconductors

We have developed an array of specialized measurement systems to test the effects of mechanical stresses on the electrical performance of superconducting materials. Extensive, advanced measurement facilities are available, including high-field (18.5 teslas) and split-pair magnets, servohydraulic mechanical testing systems, and state-of-the-art measurement probes. These probes are used for research on the effects of axial tensile strain and transverse compressive strain on critical current, measurement of cryogenic stress-strain characteristics, composite magnetic coil testing, and variable-temperature magnetoresistance measurements. Our

Technical Contact:
Jack Ekin

Staff (FY 2006):
1.0 professional
0.5 technician
2.0 research associates

electromechanical test capability for superconductors is one of only a few in the world.

COLLABORATION WITH OTHER GOVERNMENT AGENCIES

These measurements are an important element of our ongoing work with the U.S. Department of Energy (**DOE**). The DOE Office of High Energy Physics sponsors our research on electromechanical properties of candidate superconductors for particle-accelerator magnets. These materials include low-temperature superconductors (Nb_3Sn, Nb_3Al, and MgB_2), and high-temperature superconductors — Bi-Sr-Ca-Cu-O (**BSCCO**) and Y-Ba-Cu-O (**YBCO**) — including conductors made on rolling-assisted, biaxially textured substrates (**RABiTS**) and conductors made by ion-beam-assisted deposition (**IBAD**). Our research is also sponsored by the DOE Office of Electric Transmission and Distribution. Here, we focus on high-temperature superconductors for power applications, including power-conditioning systems, motors and generators, transformers, magnetic energy storage, and transmission lines. In all these applications, the electromechanical properties of these inherently brittle materials play an important role in determining their successful utilization.

CHARACTERIZATION OF SUPERCONDUCTORS FOR ELECTRIC POWER GRID RELIABILITY

Improved superconductors are being developed by U.S. companies and demonstrated for power transmission. Superconductors' greater current carrying capability is advantageous for upgrading real-estate-limited transmission lines in cities. Superconductors are also being developed for use in superconductor magnetic energy storage (**SMES**). Our work on characterizing superconducting properties at high stresses and high strains, and over variable temperatures is critical for the development of these superconductors. This work is also supported by DOE.

Significant progress in second-generation superconductors was reported in 2005-2006. These thin, highly textured YBCO films are deposited with mainly non-vacuum techniques on flexible metal substrates. They are now available in lengths of over 300 meters, carrying very high currents of over 2.5 to 3.0 mega-amperes per square centimeter at 77 kelvins. These superconductors have the potential to replace and improve parts of the ageing power grid in the United States. However, with the first coils fabricated from second-generation conductors, manufacturers learned that the layered architecture of the conductor may pose a problem: delamination of the ceramic layers under transverse tensile stress. This is important for rotating machinery, because of the centrifugal forces on the conductors, and more generally, because differential contraction in coil structures can place the conductors under severe transverse tensile stresses.

SCALING LAWS FOR MAGNET DESIGN

In the area of low-temperature superconductors, we are generalizing the Strain Scaling Law (**SSL**), a magnet design relationship we discovered two decades ago. Since then, the SSL has been used in the structural design of most large magnets, based on superconductors with the A-15 crystal structure. However, this relationship is a one-dimensional law. We are developing a measurement system to carefully determine the three-dimensional strain effects in A-15 superconductors. The importance of these measurements for very large accelerator magnets is considerable. The SSL is also being developed for high-temperature superconductors, since we recently discovered that practical high-temperature superconductors also exhibit an intrinsic axial-strain effect.

ACCOMPLISHMENTS

■ **New Apparatus Developed to Measure Delamination in YBCO Coated Conductors** — Over the past two years we have measured the delamination strength of second-generation YBCO superconductors. This required the design and construction of a new test apparatus. The test fixture head of the apparatus consists of two anvils made of Ni-5at.%W, which is the same material as the substrate of most coated conductors. The bottom anvil is soldered to the substrate side of the sample, while the top anvil is soldered to the silver cap layer on top of the ceramic YBCO layer. The choice of fabricating both anvils from the same material as the substrate of the superconductor ensures the absence of thermal shear stress between the sample and the anvils when they are cooled to 77 kelvins. The transverse tensile strength of the conductor is measured in two steps. First, the internal strength of the ceramic layers is measured, without edge effects. Second, the overall strength is measured, including edge effects. Edge damage may arise from, for instance, slitting of the conductor to smaller widths (a common procedure in the manufacturing process), and we anticipated that it may play a role in initiating delamination.

Test fixture head of a new apparatus to measure the delamination strength of second-generation high-temperature superconductors. The photo shows a YBCO coated conductor tape mounted between the two anvils of the apparatus.

Indeed, successful testing with the new apparatus showed that slit conductors have a relatively low transverse tensile strength. The overall strength of the conductor is reduced significantly by the slitting process. One company developed a structure to reinforce the slit conductor by soldering copper strips around the conductor (three-ply structure). The added solder joints at the edges of the conductor help restore the overall strength. We are currently collaborating with another company to find other novel solutions to avoid edge damage caused by slitting. In particular, we are pursuing solutions that produce a strong conductor without the need for extra reinforcement that lowers the conductor's overall critical current density.

■ **Discovery of Large Universal Effect of Axial Strain on the Critical Current of High-Temperature Superconductors** — Although remarkable technical advances have been achieved during the past several years in the development of high-temperature superconductors (**HTS**) for use in large-scale applications, these have occurred without a clear understanding of the underlying mechanism of superconductivity in these materials. One of the main areas not fully understood is the change of the superconducting current density with applied strain. We have discovered a very large universal, reversible change in the J_c of YBCO coated conductors, which is symmetric under both high compressive and high tensile strain. We anticipate these findings will initiate detailed research on the effect of strain on the underlying mechanisms of superconductivity in practical HTS. For instance, strain fields at grain boundaries in HTS may be the main limiting mechanism for supercurrents.

The superconducting current density in its self-field decreases reversibly by more than 40 percent under compressive strain in a wide range of conductors fabricated by vastly different processes. The effect is nearly the same for all high-current conductors measured, including YBCO deposited by metal-organic chemical vapor deposition (**MOCVD**) (which results in a columnar YBCO grain structure), metal-organic deposition (**MOD**) (which results in a laminar YBCO grain structure), and a hybrid conductor fabricated with a double YBCO layer (consisting of YBCO that is doped with a large amount of Dy particles that provide extra flux pinning centers). The effect is symmetric under both compressive and tensile strains.

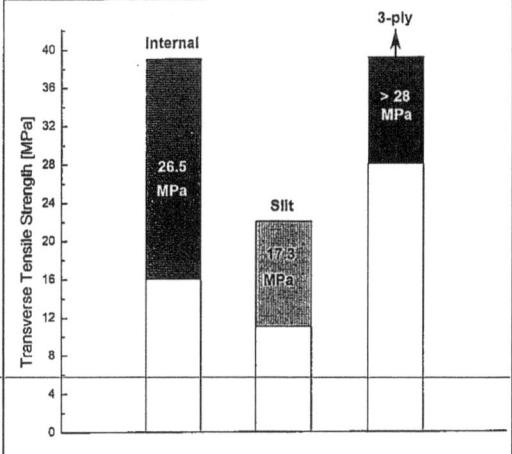

Conductor slitting reduces the overall transverse tensile strength to an average of 17.3 megapascals (numbers in the bars indicate the average value), which is far below the average internal strength of 26.5 megapascals. Reinforcing the slit sample by soldering copper strips (three-ply structure) raised the average transverse tensile strength to 24.8 megapascals. The solid parts of the bars give an indication of the spread in strength among samples.

A major reversible strain effect is a consistent and intrinsic phenomenon of high-current YBCO biaxially aligned coated conductors, encompassing a number of different types fabricated with widely varying YBCO deposition techniques. The universality and symmetry provide evidence that the mechanism behind the reversible strain effect is an intrinsic feature of the YBCO grain structure (columnar for MOCVD-IBAD and laminar for MOD-RABiTS). The results open the door for detailed studies into the mechanism of superconductivity in HTS.

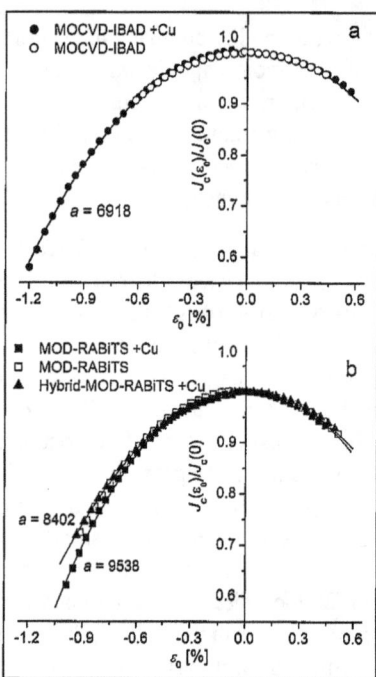

Normalized superconducting current density plotted as function of intrinsic strain ε_0 for three different types of samples: MOCVD-IBAD, MOD-RABiTS, and (hybrid) MOD-RABiTS, for both bare samples and those with copper added for stability. The solid lines describe a power-law function. The values of the strain-sensitivity parameter a are included in the figure. Compressive strain is indicated by negative values of ε_0; tensile strain is positive.

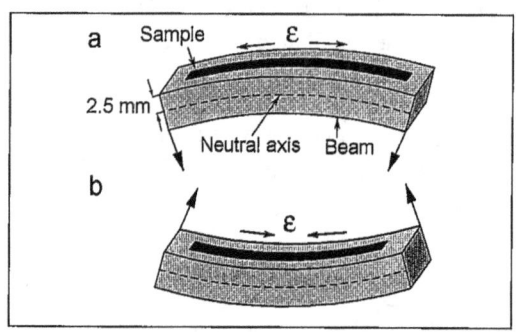

Illustration of method for applying large axial strains to superconducting sample that is soldered on top of a Cu-2%Be bending beam. Axial tension is applied by bending the beam in the direction shown in (a), whereas axial compression is applied by bending the beam in the opposite direction (b). Strain is uniform over the thickness of the superconducting films to within about 1 part in 2500.

■ **Slit YBCO Coated Conductors Prove Mechanically Robust Under Fatigue Cycling** — In order to evaluate the effect of slitting, we used fatigue cycling under transverse compressive stress, since earlier experiments had shown this to induce crack propagation. Stress was applied to the tape sample by means of two stainless-steel anvils. Uniformity of stress over the pressed area of the conductor was achieved by beveling the edges of the top anvil, and attaching it to a biaxially gimbaled pressure foot so that this anvil conforms precisely to the bottom anvil and sample surfaces. Stress was cycled between positive and negative 150 megapascals (about twice the stress level of most applications) at a frequency of 1 hertz for up to 20,000 fatigue cycles. J_c was measured at 76 kelvins in self-field. These tests simulate conditions in applications such as electric generators and industrial magnets, and evaluate whether fatigue exacerbates cracks by propagating them into the middle of the conductor. The measurements also help discriminate between different slitting techniques. The samples investigated exhibited no significant degradation under fatigue testing, thus demonstrating that, unlike transverse-tension testing, slitting does not affect the sample performance under transverse-compressive-stress cycling.

Fatigue test fixture showing the top anvil, biaxially gimbaled to uniformly apply pressure to the conductor.

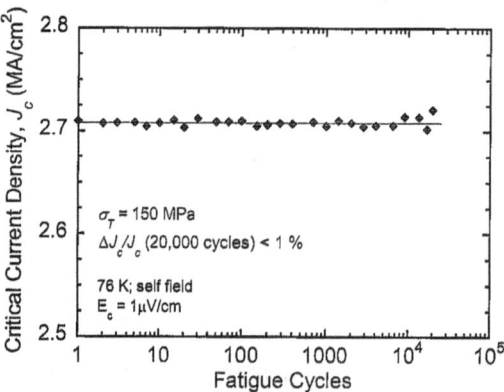

Effect of fatigue cycling under transverse compressive stress in a YBCO RABiTS sample laminated with Cu foils on both sides (three-ply geometry). J_c showed no significant degradation under fatigue testing up to 20,000 cycles at 150 megapascals.

Book

Jack Ekin's new textbook, *Experimental Techniques for Low-Temperature Measurements*, was published by Oxford University Press in October 2006. The text covers the design of cryogenic measurement probes, and the appendix provides cryogenic materials data for carrying out that design. The textbook is written for beginning graduate students, industry measurement engineers, and materials scientists interested in learning how to design successful low-temperature measurement systems. Topics include heat-transfer techniques for designing a cryogenic apparatus, selecting materials appropriate for such apparatus, how to make high-quality electrical contacts to a superconductor, and techniques for reliable critical-current measurements.

The appendix is a data handbook of materials properties and cryostat design consisting of 70 tables compiled from over 50 years of literature. The tables were compiled for experts in the field of cryogenic measurements and include electrical, thermal, magnetic, and mechanical properties of materials for cryostat construction; properties of cryogenic liquids; and temperature measurement tables and thermometer properties.

Cover of new book on cryogenic measurement techniques.

Award

Superconductivity for Electric Systems Program, Office of Electricity Delivery and Energy Reliability, U.S. Department of Energy, recognition as top ranked project, 2005 and 2006 (Jack Ekin, Najib Cheggour, and Danko van der Laan).

Recent Publications

D. C. van der Laan and J. W. Ekin, "Large Intrinsic Effect of Axial Strain on the Critical Current of High-Temperature Superconductors for Electric Power Applications" (submitted for publication).

N. Cheggour, J. W. Ekin, C. L. H. Thieme, and Y.-Y. Xie, "Effect of Fatigue Under Transverse Compressive Stress in Slitted Y-Ba-Cu-O Coated Conductors" (submitted for publication).

N. Cheggour, J. W. Ekin, and L. F. Goodrich, "Critical-Current Measurements on ITER Nb_3Sn Strands: Effect of Axial Tensile Strain" (submitted for publication).

L. F. Goodrich, N. Cheggour, J. W. Ekin, and T. C. Stauffer, "Critical-Current Measurements on ITER Nb_3Sn Strands: Effect of Temperature" (submitted for publication).

J. W. Ekin, *Experimental Techniques for Low-Temperature Measurements — Cryostat Design, Material Properties and Superconductor Critical-Current Testing*, Oxford Univ. Press, U.K. (October 2006).

C. C. Clickner, J. W. Ekin, N. Cheggour, C. L. H. Thieme, Y. Qiao, Y.-Y. Xie, and A. Goyal. "Mechanical Properties of Pure Ni and Ni-Alloy Substrate Materials for Y Ba Cu O Coated Conductors," *Cryogenics* **46**, 432-438 (June 2006).

D. C. van der Laan, J. W. Ekin, H. J. N. van Eck, M. Dhalle, B. ten Haken, M. W. Davidson, and J. Schwartz, "Effect of Tensile Strain on Grain Connectivity and Flux Pinning in $Bi_2Sr_2Ca_2Cu_3O_x$ Tapes," *Appl. Phys. Lett.* **88**, 022511 (January 2006).

N. Cheggour, J. W. Ekin, C. L. H. Thieme, Y.-Y. Xie, V. Selvamanickam, and R. Feenstra. "Reversible Axial-Strain Effect in Y Ba-Cu-O Coated Conductors," *Supercond. Sci. Technol.* **18**, S319-S324 (December 2005).

N. Cheggour, J. W. Ekin, Y.-Y. Xie, V. Selvamanickam, C. L. H. Thieme, and D. T. Verebelyi, "Enhancement of the Irreversible Axial-Strain Limit of Y Ba-Cu-O Coated Conductors with the Addition of a Cu Layer," *Appl. Phys. Lett.* **87**, 212505 (November 2005).

Y.-Y. Xie, A. Knoll, Y. Chen, Y. Li, X. Xiong, Y. Qiao, P. Hou, J. Reeves, T. Salagaj, K. Lenseth, L. Civale, B. Maiorov, Y. Iwasa, V. Solovyov, M. Suenaga, N. Cheggour, C. Clickner, J. W. Ekin, C. Weber, and V. Selvamanickam. "Progress in Scale-Up of Second-Generation High-Temperature Superconductors at Superpower," *Physica C* **426**, 849-857 (October 2005).

N. Cheggour, J. W. Ekin, and C. L. H. Thieme, "Magnetic-Field Dependence of the Reversible Axial-Strain Effect in Y-Ba-Cu-O Coated Conductors," *IEEE Trans. Appl. Supercond.* **15**, 3577-3580 (June 2005).

J. W. Ekin, N. Cheggour, M. Abrecht, C. Clickner, M. Field, S. Hong, J. Parrell, and Y.-Z. Zhang, "Compressive Pre-Strain in High-Niobium-Fraction Nb_3Sn Superconductors," *IEEE Trans. Appl. Supercond.* **15**, 3560–3563 (June 2005).

N. N. Martovetsky, P. Bruzzone, B. Stepanov, R. Wesche, C.-Y. Gung, J. V. Minervini, M. Takayasu, L. F. Goodrich, J. W. Ekin, and A. Nijhuis, "Effect of Conduit Material on CWX Performance Under High Cycling Loads," *IEEE Trans. Appl. Supercond.* **15**, 1367-1370 (June 2005).

J. W. Ekin, "Superconductors: An Emerging Power Technology," 10th Gaseous Dielectrics Conference, L. G. Chrisophorou, J. K. Olthoff, and P. Vassiliou, Eds., Springer-Verlag, Berlin, pp. 423–432 (2004).

Y.-Z. Xu and J. W. Ekin, "Tunneling Characteristics and Low-Frequency Noise of High-T_c Superconductor/Noble-Metal Junctions," *Phys. Rev. B* **69**, 104515 (March 2004).

STANDARDS FOR SUPERCONDUCTOR AND MAGNETIC MEASUREMENTS

GOALS

This project develops standard measurement techniques for critical current, residual resistivity ratio, and magnetic hysteresis loss, and provides quality assurance and reference data for commercial high-temperature and low-temperature superconductors. Applications supported include magnetic-resonance imaging, research magnets, magnets for fusion confinement, motors, generators, transformers, high-quality-factor resonant cavities for particle accelerators, and superconducting bearings. Superconductor applications specific to the electrical power industry include transmission lines, synchronous condensers, magnetic energy storage, and fault-current limiters. Project members assist in the creation and management of international standards through the International Electrotechnical Commission for superconductor characterization covering all commercial applications, including electronics. The project is currently focusing on measurements of variable-temperature critical current, residual resistivity ratio, magnetic hysteresis loss, critical current of marginally stable superconductors, and the irreversible effects of changes in magnetic field and temperature on critical current.

Probe for the measurement of the critical current of a superconductor wire as a function of temperature. The probe is inserted into the bore of a high field superconducting magnet.

CUSTOMER NEEDS

This project serves the U.S. superconductor industry, which consists of many small companies, in the development of new metrology and standards, and in providing difficult and unique measurements. We participate in projects sponsored by other government agencies that involve industry, universities, and national laboratories.

The potential impact of superconductivity on electric power systems, alternative energy sources, and research magnets makes this technology especially important. We focus on: (1) developing new metrology needed for evolving, large-scale superconductors, (2) providing unique databases of superconductor properties, (3) participating in interlaboratory comparisons needed to verify techniques and systems used by U.S. industry, and (4) developing international standards for superconductivity needed for fair and open competition and improved communication.

Electric power grid stability, power quality, and urban power needs are pressing national problems. Superconductive applications can address many of them in ways and with efficiencies that conventional materials cannot. "Second-generation" Y-Ba-Cu-O (**YBCO**) superconductors are approaching the targets established by the U.S. Department of Energy. The demonstration of a superconductor synchronous condenser for reactive power support was very successful and has drawn attention to the promise of this technology. Previous demonstration projects had involved first-generation materials, Bi-Sr-Ca-Cu-O (**BSCCO**). Variable-temperature measurements of critical current and magnetic hysteresis loss will become more important with these AC applications, and methods for reducing losses are expected to evolve as second-generation materials become commercial.

Fusion energy is a potential, virtually inexhaustible energy source for the future. It does not produce CO_2 and is environmentally cleaner than fission energy. Superconductors are used to generate the ultrahigh magnetic fields that confine the plasma in fusion energy research. We measure the magnetic hysteresis loss and critical current of marginally stable, high-current Nb_3Sn superconductors for fusion and other research magnets. Because of the way superconductors are used in magnets, *variable-temperature* critical-current measurements are needed for more complete characterization.

The focus of high-energy research is to probe and understand nature at the most basic level, including dark matter and dark energy. The particle accelerator and detector magnets needed for this fundamental science continue to push the limits of superconductor technology. The next generation of Nb_3Sn and Nb-Ti wires is pushing towards higher current density, less stabilizer, larger wire diameter, and higher magnetic fields. The resulting higher

Technical Contact:
Loren Goodrich

Staff-Years (FY 2006):
1.0 professional
0.7 technician

current required for critical-current measurements turns many minor measurement problems into significant engineering challenges. For example, heating of the specimen, from many sources, during the measurement can cause a wire to appear to be thermally unstable. Newer MgB_2 wires may be used for specialty magnets that can safely operate at the higher temperatures caused by high heat loads. We need to make sure that our measurements and the measurements of other laboratories keep up with these challenges and provide accurate results for conductor development, evaluation, and application.

Possible spin-off applications of particle accelerators are efficient, powerful light sources and free-electron lasers for biomedicine and nanoscale materials production. The heart of these applications is a linear accelerator that uses high-efficiency, pure Nb resonant cavities. We conduct research on a key materials property measurement for this application, the residual resistivity ratio (**RRR**) of the pure Nb. This measurement is difficult because it is performed on samples that have dimensions similar to those of the application. Precise variable-temperature measurements are needed for accurate extrapolations.

Technical Strategy

International Standards

With each significant advance in superconductor technology, new procedures, interlaboratory comparisons, and standards are needed. International standards for superconductivity are created through the International Electrotechnical Commission (**IEC**), Technical Committee 90 (**TC 90**).

Critical Current Measurements

One of the most important performance parameters for large-scale superconductor applications is the critical current. Critical current is difficult to measure correctly and accurately; thus these measurements are often subject to scrutiny and debate. The critical current is determined from a measurements of voltage versus current. Typical criteria are electric-field strength of 10 microvolts per meter and resistivity of 10^{-14} ohm-meters.

Critical-current measurements at variable temperatures are needed to determine the temperature margin for magnet applications. The temperature margin is defined as the difference between the operating temperature and the temperature at which critical current I_c is equal to the operating current. When a magnet is operating, transient excursions in magnetic field H or current I are not expected; however, many events can cause transient excursions to higher temperatures T, such as wire motion, AC losses, and radiation. Hence the temperature margin of a wire is a key specification in the design of superconducting magnets. Variable-temperature critical-current measurements require data acquisition with the sample in a flowing gas environment rather than immersed in a liquid cryogen. Accurate high-current (above 100 amperes) measurements in a flowing gas environment are very difficult to perform.

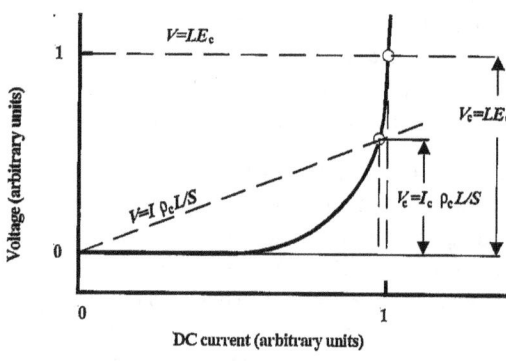

Illustration of a superconductor's voltage-current characteristic with two common criteria applied.

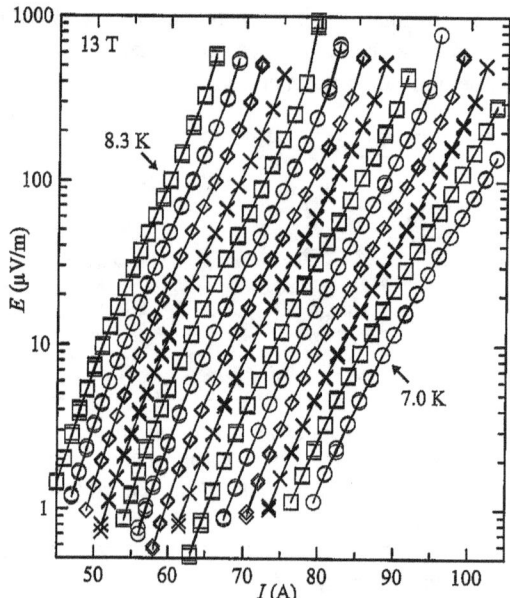

Electric field versus current at temperatures from 7.0 to 8.3 kelvins in steps of 0.1 kelvins for a Nb_3Sn wire. These are typical curves for the determination of critical current.

Residual Resistivity Ratio Measurements

The *RRR* is defined as the ratio of electrical resistivity at two temperatures: 273 kelvins (0 degrees

Celsius) and 4.2 kelvins (the boiling point of liquid helium). The value of *RRR* indicates the purity and the low-temperature thermal conductivity of a material, and is often used as a materials specification for superconductors. The low temperature resistivity of a sample that contains a superconductor is defined at a temperature just above the transition temperature or is defined as the normal-state value extrapolated to 4.2 kelvins. For a composite superconducting wire, *RRR* is an indicator of the quality of the stabilizer, which is usually copper or aluminum that provides electrical and thermal conduction during conditions where the local superconductor momentarily enters the normal state. For pure Nb used in radio-frequency cavities of linear accelerators, the low temperature resistivity is defined as the normal-state value extrapolated to 4.2 kelvins. This extrapolation requires precise measurements. We have studied some fundamental questions concerning the best measurement of *RRR* and the relative differences associated with different measurement methods, model equations for the extrapolation, and magnetic field orientations (when a field is used to drive the superconductor into the normal state).

MAGNETIC HYSTERESIS LOSS MEASUREMENTS

As part of our program to characterize superconductors, we measure the magnetic hysteresis loss of marginally stable, high-current Nb_3Sn superconductors for fusion and particle-accelerator magnets. We use a magnetometer based on a superconducting quantum interference device (**SQUID**) to measure the magnetic hysteresis loss of superconductors, which is the area of the magnetization-versus-field loop. In some cases, especially for marginally stable conductors, we use special techniques to obtain accurate results. Measurement techniques developed at NIST have been adopted by other laboratories.

ACCOMPLISHMENTS

- **Superconductor Data Enables U.S. Company to Offer Product to Korean Project** — New bismuth-based high-temperature superconductor wires are under active consideration for a 600 kilojoule superconducting magnetic energy storage (**SMES**) project lead by the Korea Electrotechnology Research Institute. The purpose of the SMES system is to stabilize the electric power grid. The magnet will be wound with 10-kilo-ampere superconducting cables composed of many round wires. It will be cooled to 20 kelvins by cryocoolers.

A U.S. company turned to us for critical current measurements at 20 kelvins to determine whether its conductor could meet the project's specifications for critical current. Critical current, the largest current a superconducting wire can carry, is a key performance and design parameter. Critical current depends on temperature, magnetic field, and, in many cases, the angle of the magnetic field with respect to the conductor.

We made variable-temperature critical-current measurements on three wire specimens in magnetic fields up to 8 teslas, at various magnetic-field angles, and at temperatures from 4 to 30 kelvins. NIST has the only such multiparameter, high-current, variable-temperature measurement capability in the U.S. The largest current applied to the 0.81 millimeter diameter wire samples was 775 amperes.

The results showed that the angle dependence of critical current for the wires was less than just 3 percent over the useful range of field and temperature, and that the round wires could be used at higher magnetic fields and temperatures than tape conductors. These data will be used to design the safe operating limits of the SMES magnet system.

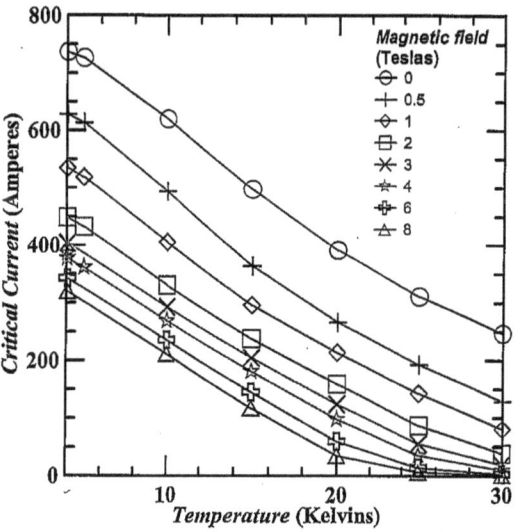

Critical current versus temperature of a high-T_c $Bi_2Sr_2CaCu_2O_{8+x}$ wire at various magnetic fields. Such curves are used to determine the safe operating current at different temperatures and fields.

- **Key Measurements for the International Thermonuclear Experimental Reactor** — Superconducting magnets are used in fusion energy projects, such as the International Thermonuclear Experimental Reactor (**ITER**), to confine and heat the plasma. The superconductors for ITER's large magnet systems are all "cable-in-conduit conductors" (**CICC**), which provide both mechanical support for the large magnetic forces and a flow path for the liquid helium required to cool the cable. The superconducting magnet must be operated below the critical current of the cable, which is a function of magnetic field and temperature. Temperature is an important variable, and the local temperature of the conductor depends on the mass-flow rate of the coolant and the distribution of the heat load along the CICC.

We designed and constructed a new variable temperature probe that allows us to make measurements in our 52-millimeter bore, 16-tesla magnet. This probe replaces one that was designed for our 86-millimeter bore, 12-tesla magnet. Fitting everything into the smaller bore was difficult, but the new probe performed as expected and allows us to make measurements at the ITER design field of 13 teslas. We made measurements up to 765 amperes with a Nb_3Sn sample in flowing helium gas. Measurements were made at temperatures from 4 to 17 kelvins and magnetic fields from 0 to 14 teslas. Some measurements were made at 15 and 16 teslas for temperatures from 4 to 5 kelvins; however, these magnetic fields can be generated only when a sample is measured in liquid helium. The results of our unique variable-temperature measurements provide a comprehensive characterization and form a basis for evaluating CICC and magnet performance. We used these data to

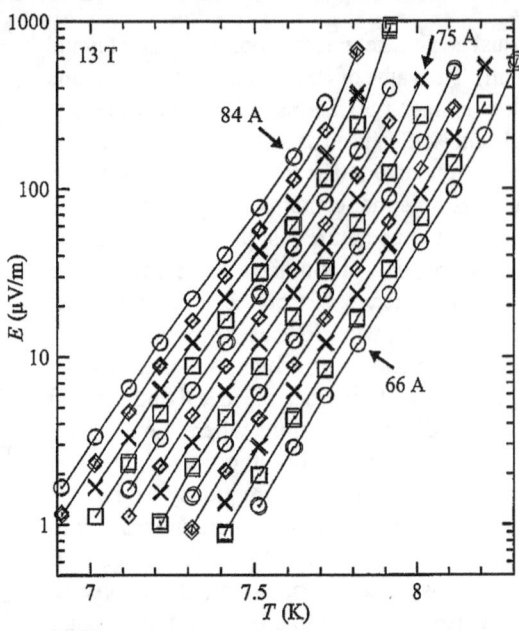

Electric field versus temperature at currents from 66 to 84 amperes in steps of 1.5 amperes for a Nb_3Sn wire. These are typical curves for the determination of temperature margin.

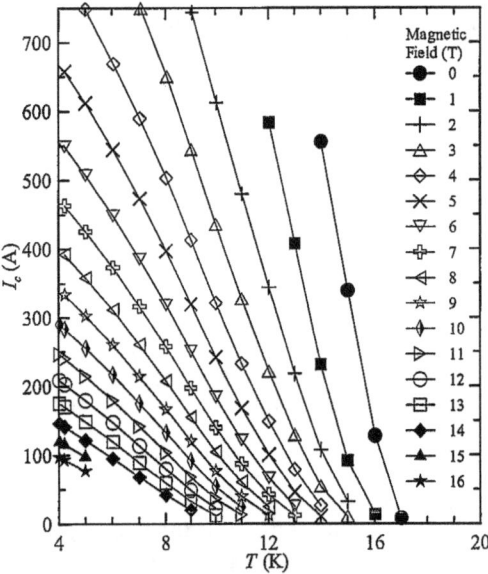

Critical current versus temperature at various magnetic fields for a Nb_3Sn wire. These curves show the current carrying limits for various combinations of temperature and magnetic field.

generate curves of electric field versus temperature at constant current and magnetic field. In turn, these give a direct indication of the temperature safety margin of the conductor.

■ **International Standards on Superconductivity** — Many of the 14 published IEC/TC 90 standards on superconductivity contain "precision" and "accuracy" statements rather than currently accepted statements of "uncertainty." NIST has advocated that TC 90 adopt a more modern approach to uncertainty. In collaboration with the Information Technology Laboratory, we have developed a 50-page report on the possibility of changing statements of "accuracy" to statements of "uncertainty" in IEC/TC 90 measurement standards, which was presented at TC 90 meetings in June 2006. They included proposed change sheets for 13 of the 14 TC 90 document standards. Ultimately, all TC 90 delegates voted in favor of changing to uncertainty statements during the maintenance cycle of existing standards and during the development of new standards.

■ **Current Ripple a Source of Measurement Errors** — All high-current power supplies contain some current ripple and spikes. New high-performance conductors have high critical currents that require current supplies over 1000 amperes. High-current power supplies with the lowest level of current ripple and spikes are often more than a factor of ten times more expensive than conventional supplies. In addition, current ripple and spikes are a greater problem for short-sample critical current testing than for magnet operation because of the smaller load inductance. Therefore, we need to understand the effects of ripple and spikes on the measured critical current (I_c) and "***n*-value**," the index of the shape of the electric field-current curve. We focused on how ripple changes the *n*-value and showed that, in terms of percentage change, the effect of ripple on *n*-value was about 7 times that on I_c. Interlaboratory comparisons often show variations in *n*-value much larger than the variations in I_c. We examined models and use the measurements on simulators to attempt to reproduce and understand the effects observed in measurements on superconductors. We believe that current ripple and spikes are sources of differences in *n*-values measured at different laboratories.

- **New Method to Evaluate the Relative Stability of Conductors** — We recently started measuring voltage versus magnetic field (***V-H***) on Nb_3Sn wires to assess their relative stability. Voltage versus current (***V-I***) at constant field is usually measured to determine I_c. Low-noise *V-H* measurements were made at constant or ramping current with the same electronic instruments, apparatus, and sample mount as used in I_c measurements. High-performance Nb_3Sn wires exhibit flux-jump instabilities at low magnetic fields, and low-noise *V-H* curves on these wires show indications of flux jumps. *V-H* measurements also reveal that less stable wires will quench (abruptly and irreversibly transition to the normal state) at currents much smaller than I_c at the lower magnetic fields. This new method needs to be further understood and may be standardized to ensure that it provides accurate and reliable data.

STANDARDS COMMITTEES

- Loren Goodrich is the Chairman of IEC/TC 90, the U.S. Technical Advisor to TC 90, the Convener of Working Group 2 (WG2) in TC 90, the primary U.S. Expert to WG4, WG5, WG6 and WG11, and the secondary U.S. Expert to WG1, WG3, and WG7.

- Ted Stauffer is Administrator of the U.S. Technical Advisory Group to TC 90.

STANDARDS

In recent years, we have led in the creation and revision of several IEC standards for superconductor characterization:

- IEC 61788-1 Superconductivity – Part 1: Critical Current Measurement – DC Critical Current of Cu/Nb-Ti Composite Superconductors

- IEC 61788-2 Superconductivity – Part 2: Critical Current Measurement – DC Critical Current of Nb_3Sn Composite Superconductors

- IEC 61788-3 Superconductivity – Part 3: Critical Current Measurement – DC Critical Current of Ag-sheathed Bi-2212 and Bi-2223 Oxide Superconductor

- IEC 61788-4 Superconductivity – Part 4: Residual Resistance Ratio Measurement – Residual Resistance Ratio of Nb-Ti Composite Superconductors

- IEC 61788-5 Superconductivity – Part 5: Matrix to Superconductor Volume Ratio Measurement – Copper to Superconductor Volume Ratio of Cu/Nb-Ti Composite Superconductors

- IEC 61788-6 Superconductivity – Part 6: Mechanical Properties Measurement – Room Temperature Tensile Test of Cu/Nb-Ti Composite Superconductors

- IEC 61788-7 Superconductivity – Part 7: Electronic Characteristic Measurements – Surface Resistance of Superconductors at Microwave Frequencies

- IEC 61788-8 Superconductivity – Part 8: AC Loss Measurements – Total AC loss Measurement of Cu/Nb-Ti Composite Superconducting Wires Exposed to a Transverse Alternating Magnetic Field by a Pickup Coil Method

- IEC 61788-9 Superconductivity – Part 9: Measurements for bulk high temperature superconductors – Trapped flux density of large grain oxide superconductors

- IEC 61788-10 Superconductivity – Part 10: Critical Temperature Measurement – Critical Temperature of Nb-Ti, Nb_3Sn, and Bi-System Oxide Composite Superconductors by a Resistance Method

- IEC 61788-11 Superconductivity – Part 11: Residual Resistance Ratio Measurement – Residual Resistance Ratio of Nb_3Sn Composite Superconductors

- IEC 61788-12 Superconductivity – Part 12: Matrix to Superconductor Volume Ratio Measurement – Copper to Non-Copper Volume Ratio of Nb_3Sn Composite Superconducting Wires

- IEC 61788-13 Superconductivity – Part 13: AC Loss Measurements – Magnetometer Methods for Hysteresis Loss in Cu/Nb-Ti Multifilamentary Composites
- IEC 60050-815 International Electrotechnical Vocabulary – Part 815: Superconductivity

Recent Publications

L. F. Goodrich and J. D. Splett, "Current Ripple Effect on n-Value" (submitted for publication).

L. F. Goodrich, N. Cheggour, J. W. Ekin, and T. C. Stauffer, "Critical-Current Measurements on ITER Nb_3Sn Strands: Effect of Temperature" (submitted for publication).

N. Cheggour, J. W. Ekin, and L. F. Goodrich, "Critical-Current Measurements on ITER Nb_3Sn Strands: Effect of Axial Tensile Strain" (submitted for publication).

L. F. Goodrich, "Voltage Versus Magnetic Field Measurements on Nb_3Sn Wires," *Adv. Cryo. Eng. (Materials)* **52B**, 520-527 (2006).

L. F. Goodrich and T. C. Stauffer, "Variable-Temperature Critical-Current Measurements on a Nb_3Sn Wire," *IEEE Trans. Appl. Supercond.* **15**, 3356-3359 (June 2005).

L. F. Goodrich, T. C. Stauffer, J. D. Splett, and D. F. Vecchia, "Unexpected Effect of Field Angle in Magnetoresistance Measurements of High-Purity Nb," *IEEE Trans. Appl. Supercond.* **15**, 3616-3619 (June 2005).

N. N. Martovetsky, P. Bruzzone, B. Stepanov, R. Wesche, C.-Y. Gung, J. V. Minervini, M. Takayasu, L. F. Goodrich, J. W. Ekin, and A. Nijhuis, "Effect of Conduit Material on CWX Performance Under High Cycling Loads," *IEEE Trans. Appl. Supercond.* **15**, 1367-1370 (June 2005).

L. F. Goodrich, T. C. Stauffer, J. D. Splett, and D. F. Vecchia, "Measuring Residual Resistivity Ratio of High-Purity Nb," *Adv. Cryo. Eng. (Materials)* **50**, 41-48 (July 2004).

L. F. Goodrich and T. C. Stauffer, "Variable-Temperature Critical-Current Measurements on a Nb-Ti Wire," *Adv. Cryo. Eng. (Materials)* **50**, 338-345 (July 2004).

Magnetodynamics

Goals

This project develops instruments, techniques, and theory for the understanding of the high-speed response of commercially important magnetic materials. Techniques used include linear and nonlinear magneto-optics and pulsed inductive microwave magnetometry. Emphasis is on high-frequency (above 1 gigahertz), time-resolved measurements for the study of magnetization dynamics under large-field excitation. Research addresses the nature of coherence and damping in ferromagnetic systems and their effects on the fundamental limits of magnetic data storage. Research on spin-electronic systems and physics concentrates on theoretical analysis of spin-momentum-transfer oscillators. The project provides results of interest to the magnetic disk drive industry, developers of magnetic random-access memory, and the growing spin-electronics research community.

Customer Needs

Advances in magnetic information storage are vital to economic growth and U.S. competitiveness in the world market for computer products and electronic devices. Our primary customers are the magnetoelectronics industries involved in the fabrication of magnetic disk drives, magnetic sensors, and magnetic random-access memory (**MRAM**).

Data-transfer rates are increasing at 40 percent per year (30 percent from improved linear bit density, and 10 percent from greater disk rotational speed). The maximum data-transfer rate in these nanometric devices is currently 1 gigabits per second, with data-channel performance of over 3 gigahertz (in the microwave region), with corresponding magnetic switching times of less than 200 picoseconds. At these rates, a pressing need exists for an understanding of magnetization dynamics, and measurement techniques are needed to quantify the magnetic switching speeds of commercial materials.

The current laboratory demonstration record for storage density is over 30 gigabits per square centimeter (200 gigabits per square inch). How much further can longitudinal media (with in-plane magnetization) be pushed? Can perpendicular recording, patterned media with discrete data bits, or heat-assisted magnetic recording extend magnetic recording beyond the superparamagnetic limit at which magnetization becomes thermally unstable? We are developing the necessary metrology to benchmark the temporal performance of new methods of magnetic data storage.

The spin momentum transfer effect — or "spin torque" — offers new opportunities and challenges for the data storage and spintronics industries. In the commercial disk drive industry, spin torque degrades the performance of current-perpendicular-to-plane read-head components by driving unstable dynamics in the read-head sensor element. However, spin torque may also be used to fabricate nanoscale on-chip oscillators for telecommunications devices. We are developing theory to understand this effect that may be used to harness the spin momentum transfer effect for future magnetoelectronic applications.

Technical Strategy

Nanomagnetodynamics

Our aim is to identify future needs in the data-storage and other magnetoelectronic industries, develop new metrology tools, and do the experiments and modeling to provide data and theoretical underpinnings. We concentrate on two major problems in the magnetic-data-storage industry: (1) data-transfer rate, the problem of gyromagnetic effects, and the need for large damping without resorting to high magnetic fields, and (2) storage density and the problem of thermally activated reversal of magnetization. This has led to the development of instrumentation and experiments using magneto-optics and microwave circuits. Microwave coplanar waveguides are used to deliver magnetic-field pulses to materials under test. In response, a specimen's magnetization switches, but not smoothly. Rather, the magnetization vector undergoes precession. Sometimes, the magnetization can precess nonuniformly, resulting in the generation of spin waves or, in the case of small devices, incoherent rotation. We use several methods to detect the state of magnetization as a function of time. These include the following:

■ The magneto-optic Kerr effect (**MOKE**) makes use of the rotation of polarization of light upon reflection from a magnetized film. We have used MOKE with an optical microscope to measure equilibrium and nonequilibrium decay of magnetization in recording media.

Technical Contact:
Tom Silva

Staff-Years (FY 2006):
1.5 professionals
2.0 research associates

- The second-harmonic magneto-optic Kerr effect (**SH-MOKE**) is especially sensitive to surface and interface magnetization. We have used SH-MOKE for time-resolved vectorial measurements of magnetization dynamics and to demonstrate the coherent control of magnetization precession.

- In our pulsed inductive microwave magnetometer (**PIMM**), the changing magnetic state of a specimen is deduced from the change in inductance of a waveguide. This technique is fast, inexpensive, and easily transferable to industry. It may also be used as a time-domain permeameter to characterize magnetic materials. Since the development of the PIMM at NIST, similar systems have been built at several industrial research laboratories and universities.

While these instruments have immediate use for the characterization of magnetic data-storage materials, they are also powerful tools for the elucidation of magnetodynamic theory. The primary mathematical tools for the analysis of magnetic switching data are essentially phenomenological. As such, they have limited utility in aiding industry in its goal of controlling the high-speed switching properties of heads and media. We seek to provide firm theoretical foundations for the analysis of time-resolved data, with special emphasis on those theories that provide clear and unambiguous predictions that can be tested with our instruments.

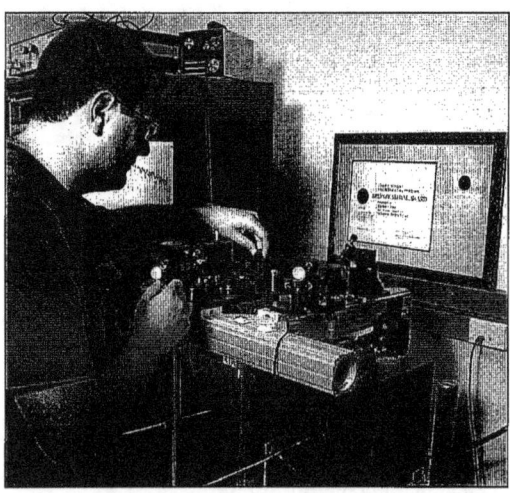

Time-resolved magneto-optic microprobe and pulsed inductive microwave magnetometer (PIMM) configured to simultaneously measure magnetization dynamics. Comparison of the two techniques permits quantitative determination of linewidth contribution due to spatial inhomogeneity of the gyromagnetic frequency.

THEORY FOR SPIN TORQUE NANO-OSCILLATORS

Our goal is to develop analytical and computational methods for the modeling and design of nanometer-scale ferromagnetic multilayer systems where the spin-momentum-transfer effect is applied. The recent discovery and rapid development of these types of systems represent major technological advances, with great promise for technological applications. Spin momentum transfer (**SMT**) generates microwave oscillations with narrow linewidths in thin magnetic multilayers. As such, these systems have the potential for next-generation signal processing and communications applications. Experimental work in this field has proceeded at an incredible rate, but theoretical understanding lags behind. Fundamental questions involving nonlinear effects on oscillator properties such as line width, power, and the frequency dependence on system parameters remain open. Our current research, in collaboration with the University of Colorado–Boulder, has demonstrated that careful mathematical modeling is very effective in describing the behavior of real systems. We have undertaken a broad investigation of SMT systems in order to fundamentally understand the SMT effect so that it may be exploited in important applications including wireless communications and fast, high-density data storage.

ACCOMPLISHMENTS

- **Analytical Model for Spin-Torque Nano-Oscillators** — We developed a nonlinear model of spin-wave excitation using a point contact in a thin ferromagnetic film. Large-amplitude magnetic solitary waves were computed using the model, which helps explain the dependence of frequency on current in recent spin-torque experiments. Numerical simulations of the fully nonlinear model predict excitation frequencies in excess of 0.2 terahertz for contact diameters smaller than 6 nanometers. These simulations also predict a saturation and red-shift of the frequency at currents large enough to invert the magnetization under the point contact. The nonlinear frequency shift caused by increasing current was found by means of numerical perturbation techniques, which agree with direct numerical simulations. The model was extended to include the Oersted magnetic field generated around the point contact due to the current flow. The presence of the Oersted field fundamentally changes the symmetry of the excited mode from even to odd symmetry for inversion about the center of the point contact.

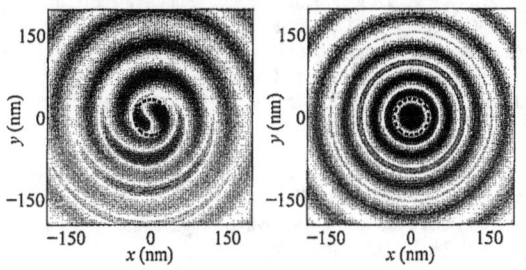

Contour plots of the x component of magnetization in a spin-torque nano-oscillator. Left: Vortex mode with odd symmetry (frequency 17.1 gigahertz); the black dashed circle in the center represents the point contact boundary. Right: Radially symmetric mode with even symmetry (frequency 18.9 gigahertz), when the Oersted field is neglected.

■ **Magnetization Dynamics Excited by Sequence of Large-Amplitude Field Pulses** — We used a time-resolved ferromagnetic resonance technique to investigate the nonlinear magnetization dynamics of a 10 nanometer thin Permalloy (Ni-Fe) film in response to a sequence of large-amplitude field pulses. The magnetic field pulse sequence was set at a repetition rate equal to the magnetic system's resonance frequency. Both inductive and optical techniques were used to observe the resultant magnetization dynamics. We compared data obtained by this technique with conventional PIMM. The results for damping and frequency response obtained by these two different methods coincide in the limit of a small-angle excitation. However, when applying large-amplitude field pulses, there was a substantial increase in the apparent damping. Analysis of vector-resolved SH-MOKE data indicate that the increase in damping is correlated with a decrease in the spatial homogeneity of the magnetization dynamics. This suggests that unstable spin-wave generation occurs in the limit of large-amplitude dynamics.

■ **Comparison of Microscopic and Spatially Averaged Magnetization Dynamics** — We adapted a time-resolved magneto-optic microprobe for use with the PIMM apparatus, allowing us to simultaneously measure the dynamics at micrometer and millimeter length scales. The microprobe has a spatial resolution of one micrometer. Comparison of the data was used to determine whether the inhomogeneous contribution to damping is in the localized or collective (that is, two-magnon-scattering) limit. These two limiting cases were theoretically delineated by the Materials Science and Engineering Laboratory. The theory establishes a minimum spacing between magnetic inhomogeneities for linewidth broadening to be considered localized in character, in analogy to inhomogeneous broadening effects for electron spin resonance and nuclear magnetic resonance. If the inhomogeneities are spaced any closer, the inhomogeneities contribute to collective excitations that are not localized at the site of the inhomogeneity. In other words, the inhomogeneities act as defects that scatter energy from the uniform mode into nearly degenerate magnon modes. Analysis of our data indicates that the inhomogeneities present in Permalloy are spaced below the limit for localization, leading us to conclude that the apparent inhomogeneous contribution to the ferromagnetic damping is actually the result of a two-magnon scattering process.

■ **Comparison of Ferromagnetic Resonance Measurements Using Different Methods** — Microwave stripline (**SL**), vector network analyzer (**VNA**), and PIMM techniques were used to measure the ferromagnetic resonance (**FMR**) linewidth for a series of Permalloy films with thicknesses of 50 and 100 nanometers. The SL-FMR measurements were made for fixed frequencies from 1.5 to 5.5 gigahertz. The VNA-FMR and PIMM measurements were made for fixed in-plane fields from 1.6 to 8 kilo-amperes per meter (20 to 100 oersteds). The results provide a confirmation, lacking until now, that the linewidths measured by these three methods are consistent and compatible. The linewidths in field are a linear function of frequency. The corresponding linewidth in frequency shows a weak upward curvature at the lowest measurement frequencies and a leveling-off at high frequencies.

COURSE

Tom Silva developed and taught a graduate level course on magnetism and magnetic materials at the University of Colorado–Boulder in spring 2006. It was attended by physicists and engineers, some from private industry.

AWARD

U.S. Department of Commerce Silver Medal (Bill Rippard, Stephen Russek, and Tom Silva) and EEEL Distinguished Associate Award (Matt Pufall and Shehzaad Kaka) for the discovery of mutual phase-locking, external frequency-locking, and frequency modulation of spin-transfer nano-oscillators, 2006.

Recent Publications

M. A. Hoefer and T. J. Silva, "Spin Momentum Transfer and Oersted Field Induce a Vortex Nano-Oscillator in Thin Ferromagnetic Film Devices" (submitted for publication).

A. B. Kos and D. C. Hurley, "Nanomechanical Mapping with Resonance Tracking Scanned Probe Microscope" (submitted for publication).

J. R. Guerrieri, M. H. Francis, P. F. Wilson, A. B. Kos, L. E. Miller, N. P. Bryner, D. W. Stroup, and L. Klein-Berndt, "RFID-Assisted Indoor Localization and Communication for First Responders," Proc. Antenna Measurement Techniques Association (AMTA), Austin, TX (October 2006); Proc. European Conference on Antennas and Propagation, Nice, France (November 2006).

S. S. Kalarickal, P. Krivosik, M. Z. Wu, C. E. Patton, M. L. Schneider, P. Kabos, T. J. Silva, and J. P. Nibarger, "Ferromagnetic Resonance Linewidth in Metallic Thin Films: Comparison of Measurement Methods," *J. Appl. Phys.* **99**, 093909 (May 2006).

T. Gerrits, M. L. Schneider, A. B. Kos, and T. J. Silva, "Large-Angle Magnetization Dynamics Measured by Time-Resolved Ferromagnetic Resonance," *Phys. Rev. B* **73**, 094454 (March 2006).

T. Gerrits, T. J. Silva, and T. Rasing, "Component-Resolved Determination of the Magnetization by Magnetization-Induced Optical Second-Harmonic Generation," *Rev. Sci. Instrum.* **77**, 034704 (March 2006).

T. Gerrits, M. L. Schneider, and T. J. Silva, "Enhanced Ferromagnetic Damping in Permalloy/Cu Bilayers," *J. Appl. Phys.* **99**, 023901 (January 2006).

M. R. Fitzsimmons, T. J. Silva, and T. M. Crawford, "Surface Oxidation of Permalloy Thin Films," *Phys. Rev. B* **73**, 014420 (January 2006).

M. A. Hoefer, M. J. Ablowitz, B. Ilan, M. R. Pufall, and T. J. Silva, "Theory of Magnetodynamics Induced by Spin Torque in Perpendicularly Magnetized Thin Films," *Phys. Rev. Lett.* **95**, 267206 (December 2005).

R. Bonin, M. L. Schneider, T. J. Silva, and J. P. Nibarger, "Dependence of Magnetization Dynamics on Magnetostriction in NiFe Alloys," *J. Appl. Phys.* **98**, 123904 (December 2005).

D. C. Hurley, M. Kopycinska-Müller, A. B. Kos, and R. H. Geiss, "Nanoscale Elastic-Property Measurements and Mapping Using Atomic Force Acoustic Microscopy Methods," *Meas. Sci. Technol.* **16**, 2167-2172 (November 2005).

S. Kaka, M. R. Pufall, W. H. Rippard, T. J. Silva, S. E. Russek, and J. A. Katine, Mutual Phase-Locking of Microwave Spin Torque Nano-Oscillators, *Nature* **437**, 389-392 (September 2005).

M. L. Schneider, T. Gerrits, A. B. Kos, and T. J. Silva, "Gyromagnetic Damping and the Role of Spin-Wave Generation in Pulsed Inductive Microwave Magnetometry," *Appl. Phys. Lett.* **87**, 072509 (August 2005).

W. H. Rippard, M. R. Pufall, S. Kaka, T. J. Silva, S. E. Russek, and J. A. Katine, "Injection Locking and Phase Control of Spin Transfer Nano-Oscillators," *Phys. Rev. Lett.* **95**, 067203 (August 2005).

D. C. Hurley, A. B. Kos, and P. Rice, "Nanoscale Elastic-Property Mapping with Contact-Resonance-Frequency AFM," in *Scanning-Probe and Other Novel Microscopies of Local Phenomena in Nanostructured Materials*, eds. S. V. Kalinin, B. Goldberg, L. M. Eng, and B. D. Huey, Proc. Mater. Res. Soc. Symp., Fall 2004, vol. 838E, paper O8.2 (June 2005).

M. L. Schneider, A. B. Kos, and T. J. Silva, "Dynamic Anisotropy of Thin Permalloy Films Measured by Use of Angle-Resolved Pulsed Inductive Microwave Magnetometry," *Appl. Phys. Lett.* **86**, 202503 (May 2005).

F. T. Charnock, R. Lopusnik, and T. J. Silva, "Pump-Probe Faraday Rotation Magnetometer Using Two Diode Lasers," *Rev. Sci. Instrum.* **76**, 056105 (May 2005).

S. E. Russek, S. Kaka, W. H. Rippard, M. R. Pufall, and T. J. Silva, "Finite-Temperature Modeling of Nanoscale Spin-Transfer Oscillators," *Phys. Rev. B* **71**, 104425 (March 2005).

M. R. Pufall, W. H. Rippard, S. Kaka, T. J. Silva, and S. E. Russek, "Frequency Modulation of Spin-Transfer Oscillators," *Appl. Phys. Lett.* **86**, 082506 (February 2005).

S. Kaka, M. R. Pufall, W. H. Rippard, T. J. Silva, S. E. Russek, J. A. Katine, and M. Carey, "Spin Transfer Switching of Spin Valve Nanopillars Using Nanosecond Pulsed Currents," *J. Magn. Magn. Mater.* **286**, 375-380 (February 2005).

J. P. Nibarger, R. L. Ewasko, M. L. Schneider, and T. J. Silva, "Dynamic and Static Magnetic Anisotropy in Thin-Film Cobalt Zirconium Tantalum," *J. Magn. Magn. Mater.* **286**, 356-361 (February 2005).

T. Gerrits, T. J. Silva, J. P. Nibarger, and T. Rasing, "Large-Angle Magnetization Dynamics Investigated by Vector-Resolved Magnetization-Induced Optical Second-Harmonic Generation," *J. Appl. Phys.* **96**, 6023-6028 (December 2004).

F. da Silva, W. C. Uhlig, A. B. Kos, S. Schima, J. Aumentado, J. Unguris, and D. P. Pappas, "Zigzag-Shaped Magnetic Sensors," *Appl. Phys. Lett.* **85**, 6022-6024 (December 2004).

W. H. Rippard, M. R. Pufall, S. Kaka, T. J. Silva, and S. E. Russek, "Current-Driven Microwave Dynamics in Magnetic Point Contacts as a Function of Applied Field Angle," *Phys. Rev. B* **70**, 100406 (September 2004).

M. L. Schneider, A. B. Kos, and T. J. Silva, "Finite Coplanar Waveguide Width Effects in Pulsed Inductive Microwave Magnetometry," *Appl. Phys. Lett.* **85**, 254-256 (July 2004).

H. Okumura, C. Y. Um, S. Y. Chu, M. E. McHenry, D. E. Laughlin, and A. B. Kos, "Structure and Magnetic Switching of Thin Film HITPERM/SiO_2 Soft Magnetic Multilayers," *IEEE Trans. Magn.* **40**, 2700-2702 (July 2004).

M. R. Pufall, W. H. Rippard, S. Kaka, S. E. Russek, T. J. Silva, J. Katine, and M. Carey, "Large-Angle, Gigahertz-Rate Random Telegraph Switching Induced by Spin-Momentum Transfer," *Phys. Rev. B* **69**, 214409 (June 2004).

W. H. Rippard, M. R. Pufall, S. Kaka, S. E. Russek, and T. J. Silva, "Direct-Current Induced Dynamics in $Co_{90}Fe_{10}/Ni_{80}Fe_{20}$ Point Contacts," *Phys. Rev. Lett.* **92**, 027201 (January 2004).

Spin Electronics

Goals

The Spin Electronics Project focuses on developing measurements to better understand the interactions between the electron spin in current-carrying electrons and the magnetization of ferromagnetic films. These measurements will allow the investigation of magnetization dynamics at nanometer lengths scales characterized by precession angles that are orders of magnitude larger than previously accessible. They will facilitate the development and continued scaling of spintronic devices. The techniques developed as part of this program will support industrial roadmaps that have targeted magnetic disk drives with terabit-per-square-inch densities and magnetic random-access memory (**MRAM**) at the 65 nanometer node by the end of the decade. In addition, these measurements will help industry develop smaller and cheaper nanoscale magnetic microwave devices that may replace much larger and expensive on-chip microwave circuitry.

Customer Needs

Until recently, the only means known to control the magnetization state of ferromagnetic structures was through the use of applied magnetic fields. However, within the last several years it has been demonstrated that this can also be accomplished through the transfer of the electron spin angular momentum from current-carrying electrons to the magnetization of magnetic films, generally referred to as the spin-momentum-transfer (**SMT**) effect. Spin transfer represents a fundamentally new way to control and manipulate the magnetic states of devices, one that emerges only at the nanoscale. It will allow investigation of magnetization dynamics and spin-waves on length scales smaller than previously possible. It creates the ability to switch nanopatterned magnetic storage devices at speeds not previously accessible. It presents opportunities for the development of spin-switched MRAM and active microwave devices operating above frequencies of 100 gigahertz.

At present, the SMT effect is too poorly understood to effectively exploit or diminish its effects in practical devices, and little is accurately known about the dynamics induced by the effect. We will develop metrology to investigate and understand how spin-based effects can be avoided or exploited in magnetic nanostructures in order to assist the magnetic data storage industry and to facilitate the development of nanoscale microwave devices operating at high frequencies. These measurements will help industry develop future generation spintronic devices and facilitate the continued scaling of magnetic data storage into the deep nanometer range.

Technical Strategy

We are performing measurements on magnetic nanostructures in order to investigate magnetization dynamics at nanometer length scales and sub-nanosecond time scales. The high-speed dynamics and switching events are induced through the spin-transfer effect, which utilizes the transfer of the electron spin angular momentum in nanometric magnetic heterostructures to induce magnetic excitations at length scales down to 10 nanometers and at frequencies up to 100 gigahertz. We are measuring the high frequency properties of the spin transfer induced dynamics in patterned magnetic nanostructures as well as dynamics that are induced locally in continuous films.

We are trying to determine how spin-based effects can be avoided and exploited in magnetic nanostructures in order to assist the magnetic data storage industry and facilitate the development of nanoscale microwave devices. We will quantify the intrinsic device properties that determine the high frequency dynamics of spin-transfer nanoscale oscillators — such as oscillation frequency, linewidth, and power, which currently only qualitatively agree with theory — and develop measurements to relate the stochastic (statistical) switching characteristics of spin-switched MRAM to the component materials properties. For example, we are presently developing techniques to measure the magnetic precessional damping parameter in single, active devices with dimensions below 100 nanometers.

Accomplishments

- **Injection Locking in Spin-Transfer Microwave Oscillators** — Synchronization of weakly coupled oscillators, generally referred to as injection locking or frequency entrainment, has numerous examples in nature and generally occurs in oscillator systems having at least weak nonlinear interactions. Examples range from biological systems, such as the synchronized flashing of fireflies and singing of certain crickets, to those in the physical sciences, such as Josephson junction arrays and the synchronization of the Moon's rota-

Technical Contact:
Bill Rippard

Staff-Years (FY 2006):
1.0 professional
2.0 research associates

tion with respect to its orbit about the Earth. This feature is exploited in many modern technologies such as wireless communications, the American power grid, various power combining architectures, and phased array antenna networks. One of the simplest methods of synchronizing electronic based oscillators is through the application of an AC signal close to the oscillator's natural frequency, inducing the device to oscillate sympathetically at the drive frequency.

We have directly measured phase locking of spin transfer oscillators to an injected AC current. In this scenario, the devices are forced to oscillate at the same frequency as the injected signal. The oscillators lock to signals up to several hundred megahertz away from their natural oscillation frequencies, depending on the relative strength of the input. As the DC current passing through the devices varies over the locking range, time-domain measurements show that the phase of the spin-transfer oscillations varies over a range of approximately plus or minus 90 degrees relative to the input. This is in good agreement with general theoretical analysis of injection locking of nonlinear oscillators.

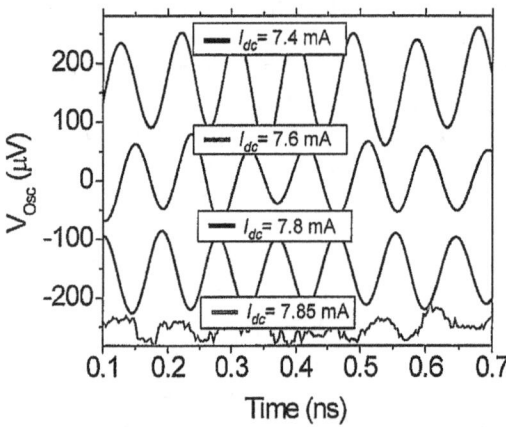

Time-domain data showing the relative variation of the oscillator phase over the locking range for several values of DC bias current.

■ **Mutual Synchronization of Spin-Transfer Oscillators** — Our demonstration that spin-transfer oscillators interact with input signals nonlinearly opens the possibility of creating phase-coherent device *arrays*. The power emitted from a single spin-torque nano-oscillator (**STNO**) is at present typically less than 1 nanowatt. To achieve a more useful power level, on the order of microwatts, a device could consist of an array of phase coherent STNOs, as has been done with arrays of Josephson junctions and larger semiconductor oscillators. We showed that two STNOs in close proximity mutu-

ally phase-lock (that is, synchronize), exhibiting a general tendency of interacting nonlinear oscillator systems.

The phase-locked state is distinct, characterized by a sudden narrowing of signal linewidth and an increase in power due to the coherence of the individual oscillators. For instance, the device oscillator linewidth decreases by a factor of 10 as compared to the individual oscillators, a significant improvement with respect to thermal fluctuations. Furthermore, we showed that the combined power output represents a fully coherent combination of the two signals, a requirement to demonstrate true phase-locking between the devices.

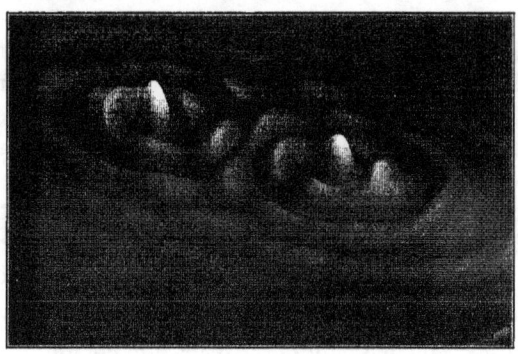

Micromagnetic simulations showing the spin-wave interaction between two local spin transfer oscillators. Each device emits spin-waves towards the other, causing the devices to synchronize.

■ **Determination of Synchronization Mechanism in Spin-Transfer Nanocontacts** — The fact that two spin-transfer nanocontact oscillators fabricated in close proximity to each other on the same magnetic film can phase-lock provides a means for coherently combining the powers of the devices. Furthermore, the power level increases as the *square* of the number of oscillators. While a promising result, an important question facing further development is the origin of the phase-locking. Spin-transfer nano-oscillators have at least two possible means of interacting. Since the devices produce magnetic fields, the oscillators could simply interact like two bar magnets, with the field from one affecting the oscillations of the other. Alternatively, since the oscillators are connected via a magnetic film, they could interact via propagating "spin-waves." Spin-waves are analogous to water waves that propagate from the splash produced by a rock thrown into a pool.

To separate these two possible mechanisms, we devised a set of experiments. First, we fabricated two oscillators on a magnetic film and showed that they interact and phase-lock as in our previous work. Then we physically cut the magnetic film between the oscillators with a focused-ion beam (**FIB**) nanoscale cutting tool. Cutting the magnetic film would still allow the two oscillators to see the magnetic fields produced by each other but would stop any spin-waves from propagating, just as a barrier dividing a pool of water stops ripples from propagating. We found that phase-locking no longer occurred, showing that the oscillators interact predominately via spin-waves.

These results set the stage for arranging larger numbers of nanocontact oscillators into coherent arrays for use in new types of electronic devices. In addition, this electrical method provides a new way to measure spin-waves themselves on length scales smaller than previously imaginable. This new metrological tool may reveal new magnetodynamic phenomena at nanometer length scales.

tive switching schemes is the scaling limitations of conventional MRAM. Spin-torque switching of MRAM has the advantage that, as the size of the devices is reduced, the current needed to switch the free layer orientation is decreased. However, one difficulty with implementing MRAM nanopillars in large-scale commercial products is device-to-device variation of the magnetic properties and the impact of these variations on the critical switching current.

To gauge device-to-device variations, we measured how thermal effects influence the free layer reversal of spin-valve nanopillars via the spin-torque effect. We found that room temperature pulsed switching probability measurements can be used to accurately predict the critical current for switching at "zero temperature." Low temperature measurements of several different devices provided experimental values of the critical switching current. Device-to-device variations in the critical switching current are drastically reduced at low temperature, with good agreement between experiment and theory.

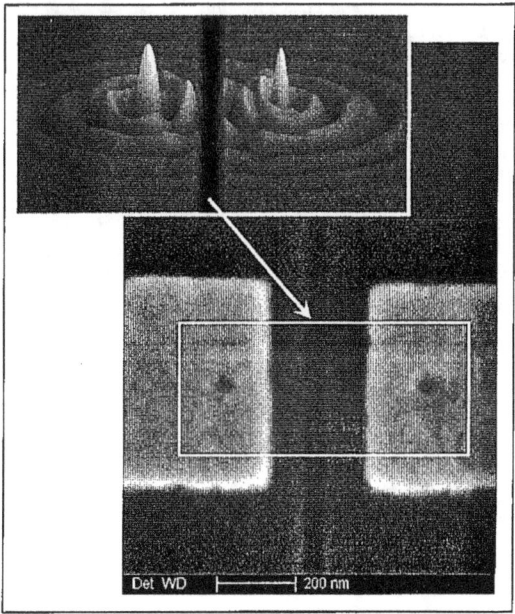

Electron micrograph of two nanocontacts (dark circles in light rectangles) with FIB cut between them. Scale bar is 200 nm. Inset: Micromagnetic simulation showing spin-waves emitted by two contacts, and reflections from FIB cut.

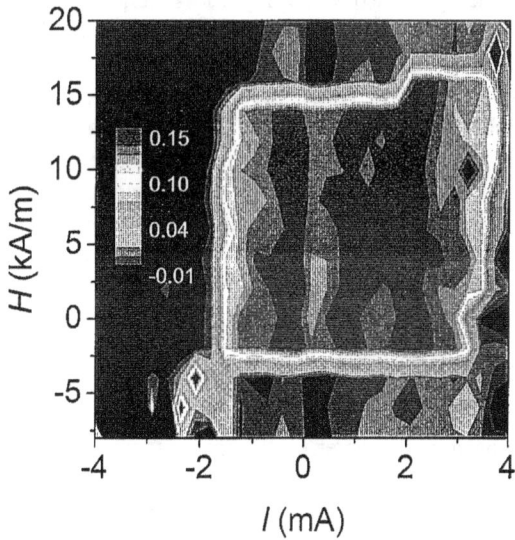

Plot of the differential resistance versus current and magnetic field for an elongated hexagon sample, 50 nanometers by 100 nanometers in size. The negative-going current sweep was subtracted from the positive-going current sweep to highlight hysteretic regions of the device.

■ **Influence of Thermal Effects in the Current-Induced Switching of Magnetic Nanostructures** — The spin-torque effect is an efficient way to change the orientation of the free layer in multilayer nanopillars known as a "spin valves." This effect has potential applications in commercial MRAM and has recently been demonstrated as a viable alternative to the cross-point writing scheme of conventional MRAM, which is just now coming to market. One of the reasons for pursuing alterna-

■ **Detailed Comparison of Spin-Transfer Precessional Dynamics with Theoretical Predictions** — Since the initial predictions that a spin polarized current can exert a torque on a nanoscale magnet, much progress has been made in understanding the spin-transfer effect and its manifestations. A number of researchers have successfully demonstrated both current-induced switching and steady-state magnetic precession in patterned magnetic nanostructures, nanowires, and nanocontacts as well as current-induced motion of magnetic domain walls. Theoretical efforts to better understand these effects have included analytical approaches, numerical single-domain modeling, and micromagnetic simulations; a rough general qualitative agreement between experiment and theory regarding the most basic results has been achieved. However, the measured precessional dynamics in magnetic nanocontacts do not always agree with the predictions of both micromagnetic and single-domain modeling. For instance, measurements show much more complicated evolution of the precessional frequency with current and field than predicted by single-domain simulations and spin-wave theories.

We have measured the detailed dependence of the oscillation frequencies, linewidths, and output powers of spin-transfer nanocontact oscillators as functions of applied field strength, bias current, and angle of the applied magnetic field. For fields applied only moderately out of the plane of the film, the evolution of these properties is continuous. However, for fields applied more strongly out of plane they exhibit discontinuities in both current and applied field. These discontinuities typically correlate with changes in the device resistance, changes in device output power, and a broadening of their spectral linewidths. However, away from these discontinuities, the oscillator output powers are larger and the linewidths narrower when compared to geometries having the fields applied at lower angles. Our measurements suggest that the discontinuous evolution of the frequency with current and applied field results from an abrupt change in precessional trajectories of the magnetization in the free layer. Comparisons with present theories give only qualitative agreement with our experimental results.

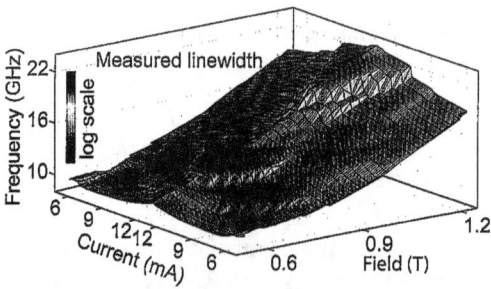

Measured data from a spin-transfer nanoscale oscillator comparing the measured device linewidth and the evolution of the precession frequency with current and magnetic field. The linewidth is shown as shading mapped directly onto the frequency surface so that the two parameters can be compared.

AWARD

U.S. Department of Commerce Silver Medal (Bill Rippard, Stephen Russek, and Tom Silva) and EEEL Distinguished Associate Award (Matt Pufall and Shehzaad Kaka) for the discovery of mutual phase-locking, external frequency-locking, and frequency modulation of spin-transfer nano-oscillators, 2006.

RECENT PUBLICATIONS

W. H. Rippard, M. R. Pufall, and S. E. Russek, "Comparison of Frequency, Linewidth, and Output Power in Spin Transfer Nanocontact Oscillators" (submitted for publication).

M. L. Schneider, M. R. Pufall, W. H. Rippard, and S. E. Russek, "Thermal Effects on the Critical Current of Spin Torque Switching in Nanopillars" (submitted for publication).

J. M. Shaw, W. H. Rippard, S. E. Russek, T. Reith, and C. M. Falco, "Narrow Switching Field Distributions in Perpendicular Magnetic Nanodot Arrays" (submitted for publication).

W. H. Rippard and M. R. Pufall, "Microwave Generation in Magnetic Multilayers and Nanostructures," in *Handbook of Magnetism and Advanced Magnetic Materials*, John Wiley, Sussex, U.K. (in press).

M. R. Pufall, W. H. Rippard, S. E. Russek, S. Kaka, and J. A. Katine, "Electrical Measurement of Spin-Wave Interactions of Proximate Spin Transfer Nanooscillators," *Phys. Rev. Lett.* **97**, 087206 (August 2006).

S. Kaka, M. R. Pufall, W. H. Rippard, T. J. Silva, S. E. Russek, and J. A. Katine, Mutual Phase-Locking of Microwave Spin Torque Nano-Oscillators, *Nature* **437**, 389-392 (September 2005).

W. H. Rippard, M. R. Pufall, S. Kaka, T. J. Silva, S. E. Russek, and J. A. Katine, "Injection Locking and Phase Control of Spin Transfer Nano-Oscillators," *Phys. Rev. Lett.* **95**, 067203 (August 2005).

S. E. Russek, S. Kaka, W. H. Rippard, M. R. Pufall, and T. J. Silva, "Finite-Temperature Modeling of Nanoscale Spin-Transfer Oscillators," *Phys. Rev. B* **71**, 104425 (March 2005).

M. R. Pufall, W. H. Rippard, S. Kaka, T. J. Silva, and S. E. Russek, "Frequency Modulation of Spin-Transfer Oscillators," *Appl. Phys. Lett.* **86**, 082506 (February 2005).

S. Kaka, M. R. Pufall, W. H. Rippard, T. J. Silva, S. E. Russek, J. A. Katine, and M. Carey, "Spin Transfer Switching of Spin Valve Nanopillars Using Nanosecond Pulsed Currents," *J. Magn. Magn. Mater.* **286**, 375-380 (February 2005).

W. H. Rippard, M. R. Pufall, S. Kaka, T. J. Silva, and S. E. Russek, "Current-Driven Microwave Dynamics in Magnetic Point Contacts as a Function of Applied Field Angle," *Phys. Rev. B* **70**, 100406 (September 2004).

M. R. Pufall, W. H. Rippard, S. Kaka, S. E. Russek, T. J. Silva, J. Katine, and M. Carey, "Large-Angle, Gigahertz-Rate Random Telegraph Switching Induced by Spin-Momentum Transfer," *Phys. Rev. B* **69**, 214409 (June 2004).

W. H. Rippard, M. R. Pufall, S. Kaka, S. E. Russek, and T. J. Silva, "Direct-Current Induced Dynamics in $Co_{90}Fe_{10}/Ni_{80}Fe_{20}$ Point Contacts," *Phys. Rev. Lett.* **92**, 027201 (January 2004).

Magnetic Devices and Nanostructures

Goals

The Magnetic Thin Films and Nanostructures Project develops measurements and standards for nanomagnetic materials and devices used in the magnetic data storage, magnetoelectronics, and biomedical industries. These measurements and standards assist industry in the development of advanced magnetic recording systems, magnetic solid-state memories, magnetic sensors, magnetic microwave devices, and biomedical materials and imaging systems. Broadband electrical measurements are being developed to characterize nanoscale devices based on giant magnetoresistance (**GMR**), tunneling magnetoresistance (**TMR**), and spin-momentum transfer (**SMT**). Magnetic resonance techniques, such as high-frequency electron paramagnetic resonance (**EPR**) and nuclear magnetic resonance (**NMR**) are being used to study the properties of nanomagnets to improve magnetic resonance imaging (**MRI**) or find applications in nanotagging. We are developing dynamic nanoscale magnetic imaging, such as time-resolved Lorentz microscopy in collaboration with the Materials Science and Engineering Laboratory, to better understand the operation of nanoscale magnetic structures and devices.

structures on nanometer-size scales and over a wide range of time scales varying from picoseconds to years. For example, the response of a 50-nanometer magnetic device, used in a read head or a magnetic random-access memory (**MRAM**) element, may be determined by a 5-nanometer region that is undergoing thermal fluctuations at frequencies of 1 hertz to 10 gigahertz. These fluctuations give rise to noise, non-ideal sensor response, and long-term memory loss. Spintronic devices and nanomagnetic materials are finding applications in other areas such as homeland security and biomedical imaging. These industries require better low-power magnetic field sensors for weapons detection, chemical detection, and magnetocardiograms, and require novel nanomagnetic materials for MRI contrast agents and defense applications.

Advances in technology are dependent on the discovery and characterization of new effects such as GMR, TMR, and SMT. Detailed understanding of spin-dependent transport is required to optimize these effects and to discover new phenomena that will lead to new spintronic device concepts. New effects such as spin momentum transfer and coherent spin transport in semiconductor devices may lead to new classes of devices that will be useful in data storage, computation, and communications applications. Many technologies require, or are enabled by, the use of magnetic nanostructures such as molecular nanomagnets. The study of magnetic nanostructures will enable data storage on the nanometer scale, a better understanding of the fundamental limits of magnetic data storage, and new biomedical applications.

Technical Strategy

We are developing several new techniques to address the needs of U.S. industries for characterization of magnetic thin films and device structures on nanometer-size scales and gigahertz frequencies.

Technical Contact:
Stephen Russek

Staff-Years (FY 2006):
1.0 professional
2.0 research associates

NMR system (7 teslas, 300-megahertz) for use in characterizing MRI contrast agents.

Customer Needs

The data storage and magnetoelectronics industries are developing smaller and faster technologies that require sub-hundred-nanometer magnetic structures to operate in the gigahertz regime. New types of spintronic devices with increased functionality and performance are being incorporated into data storage and magnetoelectronic technologies. New techniques are required to characterize these magnetic

Device Magnetodynamics

We fabricate test structures for characterizing small magnetic devices at frequencies up to 40 gigahertz. The response of submicrometer magnetic devices such as spin valves, magnetic tunnel junctions (**MTJs**), and GMR devices with current perpendicular to the plane is measured in both the linear-response and nonlinear-switching regimes. The linear-response regime is used for magnetic-recording read sensors and high-speed isolators, whereas the switching regime is used for writing or storing data in MRAM devices. We measure the sensors using microwave excitation fields and field pulses with durations down to 100 picoseconds. We compare measured data to numerical simulations of the device dynamics to determine the ability of current theory and modeling to predict the behavior of magnetic devices. We develop new techniques to control and optimize the dynamic response of magnetic devices. These include the engineering of magnetic damping by use of rare-earth doping and precessional switching, which controls switching by use of the timing of the pulses rather than pulse amplitude. This research is aimed at developing high-frequency magnetic devices for improved recording heads and for imaging of microwave currents in integrated circuits and microwave devices. Novel device structures that incorporate magnetic materials with other nanostructures, such as carbon nanotubes, are being investigated in collaboration with the Materials Science and Engineering Laboratory and the Radio-Frequency Electronics Group.

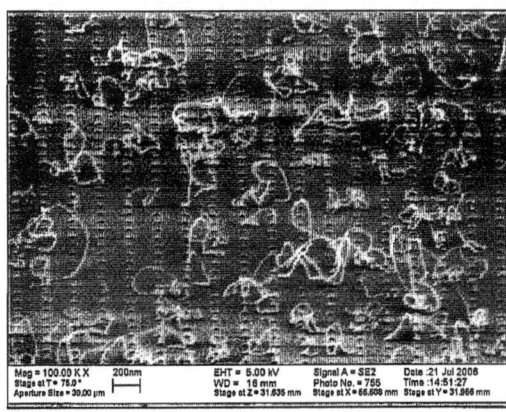

Carbon nanotubes growing from Fe nanodots for use in studying spin transport in nanostructures.

MAGNETIC NOISE AND LOW-FIELD MAGNETIC SENSORS

In collaboration with the Quantum Electrical Metrology Division, we develop new techniques to measure both the low-frequency and high-frequency noise and the effects of thermal fluctuations in small magnetic structures. Understanding the detailed effects of thermal magnetization fluctuations will be critical in determining the fundamental limit to the size of magnetic sensors, magnetic data bits, and MRAM elements. High-frequency noise is measured in our fabricated structures and in commercial read heads. High-frequency noise spectroscopy directly measures the dynamical mode structure in small magnetic devices. The technique can be extended to measure the dynamical modes in structures with dimensions as small as 20 nanometers. The stochastic motion of the magnetization during a thermally activated switching process is measured directly, which will lead to a better understanding of the long-time stability of high-density magnetic memory elements. New methods are being developed to dynamically image thermal fluctuations by use of time-resolved Lorentz and scanned probe microscopies. These new metrologies will be essential to study and control thermal fluctuations and $1/f$ noise in magnetic and spintronic devices.

NANOMAGNETISM

We are developing new methods to characterize the magnetic properties of nanomagnetic structures such as patterned media and molecular nanomagnets. Patterned magnetic nanodots are fabricated with sputter deposited magnetic multilayers and electron-beam lithography. The magnetic properties are studied with magnetic force microscopy, magneto-optical Kerr effect, and Lorentz microscopy. An important focus of this work is the ability to characterize dynamics, such as magnetic recording bit reversal, in single nanostructures at high frequencies. Magnetic nanostructures are also characterized with high-frequency EPR, based on a superconducting quantum interference device (**SQUID**) magnetometer, which can simultaneously measure low-frequency magnetic properties and high-frequency characteristics, such as resonant absorption/emission of microwaves in the frequency range of 95 to 141 gigahertz over a temperature range of 1.8 to 400 kelvins. Molecular nanomagnets, which are the smallest well defined magnetic structures that have been fabricated, exhibit quantum and thermal fluctuation effects that will necessarily be encountered as magnetic structures shrink into the nanometer regime. These systems, which contain from 3 to 12 transition-metal atoms, form small magnets with Curie temperatures of 1 to 30 kelvins. We are investigating new methods of manipulating these nanomagnets by varying the ligand structure and binding them to various films. We are looking at new applications by incorporating the nanomagnets into molecular devices and exploring how the nanomagnets relax nuclear spins in biological systems.

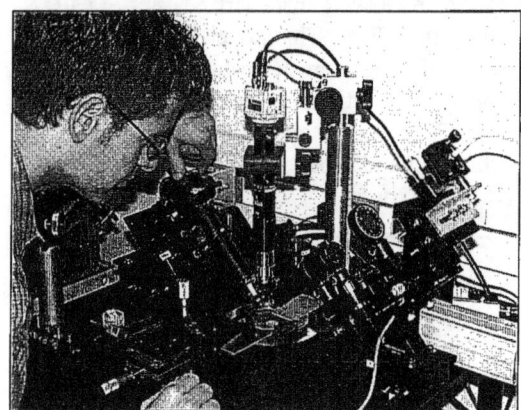

Magneto-optical Kerr effect microscope.

MRI STANDARDS AND CONTRAST AGENT METROLOGY

We develop new techniques to characterize nanomagnetic materials for improved MRI contrast and to fabricate quantitative MRI phantoms. Nanomagnetic materials in solid, liquid, or gel phases are studied with SQUID magnetometry, EPR, and NMR. We have installed an NMR relaxometer that can measure the nuclear relaxation induced in biological systems by neighboring nanomagnets. The relaxometer can measure over a field range of 0.5 to 7 teslas and NMR frequency range of 20 to 300 megahertz. Traceable standards are being developed to help monitor MRI stability and intercomparability and to enable quantitative MRI.

ACCOMPLISHMENTS

■ **Low-Frequency Noise Measurements on Magnetic Field Sensors** — Low-frequency noise was measured in the frequency range from 0.1 hertz to 10 kilohertz on a variety of commercially available magnetic sensors and custom sensors fabricated at NIST. The types of sensors investigated include those based on anisotropic magnetoresistance (**AMR**), GMR, TMR, giant magnetoimpedance (**GMI**), and fluxgate devices. The $1/f$ noise components of electronic and magnetic origin were identified by measuring sensor noise and sensitivity at various applied magnetic fields. For the GMR sensors, both electronic and magnetic components contribute to the overall sensor noise. MTJ bridge sensors, which operate on the tunneling magnetoresistance effect, were fabricated at NIST. MTJ devices consist of two magnetic layers: a free layer and a fixed layer separated by a thin insulator. The tunneling current through the insulator is dependent on the relative orientation of the two magnetic layers. These devices are used in magnetic recording read sensors, MRAM bits, and magnetic field sensors. Two different sensors are currently under development: a low power sensor (less than 100 microwatts) with a field detectivity of 1 nanotesla per root hertz was demonstrated this year, and a low field sensor with field detectivity of 1 picotesla per root hertz will be demonstrated next year. These sensors are small, low in cost, and compatible with complementary metal oxide semiconductors (**CMOS**), and will enable many applications for which there are currently no sensor solutions.

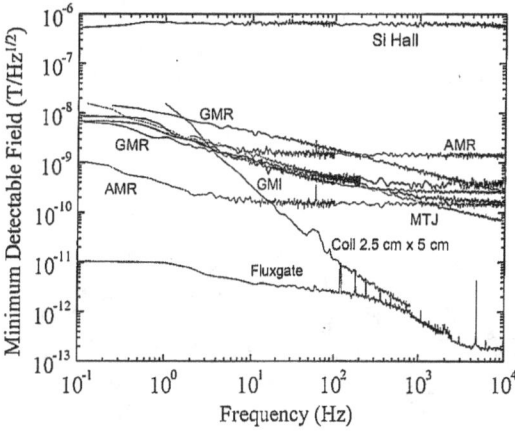

Low frequency noise measured in different types of magnetic field sensors.

■ **Dynamic Lorentz Imaging of Magnetic Tunnel Junctions** — In collaboration with the Materials Science and Engineering Laboratory, we studied disorder in MTJs using dynamic Lorentz microscopy. Lorentz microscopy is a form of transmission electron microscopy (**TEM**) that uses the deflection of electrons as they pass through a magnetic sample to image nanoscale magnetic structure. The advantages of Lorentz microscopy are that it can image the interior of a complicated device stack, it acquires data in parallel allowing rapid image acquisition needed for dynamic imaging, and it can image nanoscale structural defects that can give rise to the magnetic disorder. Considerable disorder in the free layer of the MTJ was observed during free layer rotation and switching. The disorder was a function of the MTJ preparation technique. By incorporating a nano-oxide just below the tunnel barrier (a technique developed in the Materials Science and Engineering Laboratory) the observed disorder was reduced. Time dependent fluctuations in the magnetic structure, which are a prime source of $1/f$ noise, were directly observed with dynamic Lorentz microscopy.

■ **Narrow Switching Distributions in Nanoscale Patterned Media** — Magnetic nanodots have been fabricated by electron-beam lithography to study their suitability for magnetic media. The nanodots are fabricated from multilayers of Co and Pd either using sputtering or molecular beam epitaxy. The perpendicularly magnetized nanodots have switching fields that range between 0.5 and 1.5 teslas. The switching field distributions were measured for both polycrystalline and epitaxial structures with a variety of underlayers to determine the effect of the microstructure on switching. The key results were the observation of a very narrow switching distribution in nanodots with Ta underlayers and that the epitaxial structures did not show significantly narrower switching distributions than the polycrystalline structures. This study shows that the polycrystalline nature of the nanodots, which affects both the magnetic and lithographic uniformity, is not responsible for the variation in observed switching properties.

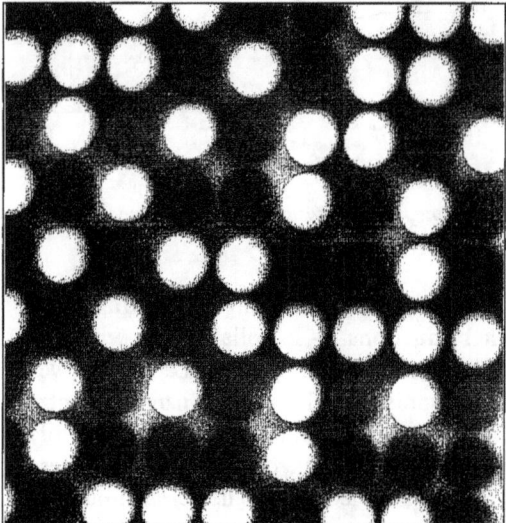

Magnetic force microscope image of 100-nanometer nanodots showing partial magnetic switching. Each nanodot is a stack of 8 layers of 0.35 nanometer of Co alternating with 8 layers of 1.02 nanometers of Pd.

■ **Evaluation of Fe_8 for use as an MRI contrast agent** — In collaboration with the University of Colorado, we completed a study of the potential use of Fe_8 molecular nanomagnets as an MRI contrast agent. NMR relaxivity data were obtained over a broad range of concentrations. At low concentrations the relaxivity, which is proportional to MRI contrast, was shown to be comparable to that of existing contrast agents. At high concentrations the relaxivity decreased. Based on a SQUID magnetometer evaluation of the decomposition of Fe_8 in aqueous solutions, the concentration dependence of the relaxivity was attributed to the concentration dependence of the decomposition rate. The excitation spectra of Fe_8 were characterized using SQUID high-frequency EPR. This novel EPR technique uses a SQUID magnetometer to quantitatively measure the spin excitation in response to microwave radiation. This work is our first effort to correlate the ESR fluctuation spectrum of a nanomagnet to the NMR relaxivity.

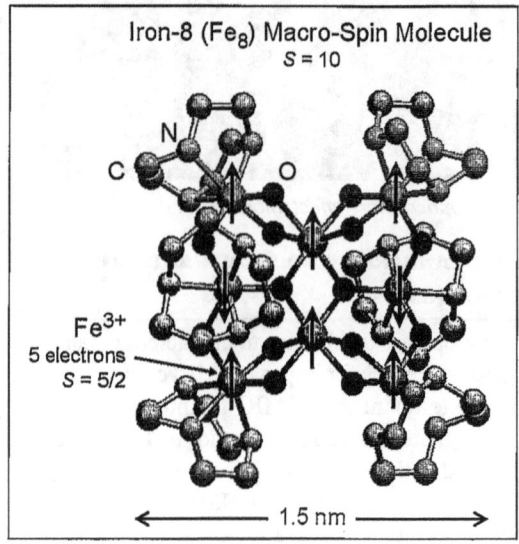

Fe_8 nanomagnets. Each Fe^{3+} ion has spin 5/2; 6 are ferromagnetically aligned and 2 are antiferromagnetically aligned. The net macro-spin is 10.

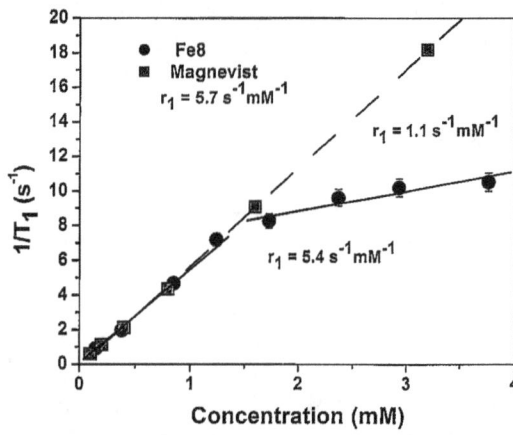

Longitudinal relaxivity of Fe_8 as a function of concentration.

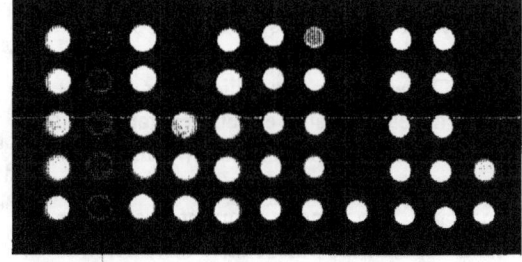

MRI phantom with multiple nanomagnetic contrast agents.

Award

U.S. Department of Commerce Silver Medal (Bill Rippard, Stephen Russek, and Tom Silva) and EEEL Distinguished Associate Award (Matt Pufall and Shehzaad Kaka) for the discovery of mutual phase-locking, external frequency-locking, and frequency modulation of spin-transfer nano-oscillators, 2006.

Recent Publications

A. McCallum, and S. E. Russek, "Refraction of Electrons at the Interfaces in Spin Valves" (submitted for publication).

J. M. Shaw, W. H. Rippard, S. E. Russek, T. Reith, and C. M. Falco, "Narrow Switching Field Distributions in Perpendicular Magnetic Nanodot Arrays" (submitted for publication).

W. H. Rippard, M. R. Pufall, and S. E. Russek, "Comparison of Frequency, Linewidth, and Output Power in Spin Transfer Nanocontact Oscillators" (submitted for publication).

M. L. Schneider, M. R. Pufall, W. H. Rippard, and S. E. Russek, "Thermal Effects on the Critical Current of Spin Torque Switching in Nanopillars" (submitted for publication).

B. Cage, S. E. Russek, R. Shoemaker, A. Barker, C. Stoldt, V. Ramachandarin, and N. Dalal, "The Utility of the Single-Molecule Magnet Fe_8 for Magnetic Resonance Imaging Contrast Agent Over a Broad Range of Concentration," *Polyhedron* (in press).

J. M. Shaw and C. M. Falco, "Structure, Spin-Dynamics, and Magnetic Properties of Annealed Nanoscale Fe Layers on GaAs," *J. Appl. Phys.* (in press).

E. Mirowski, J. Moreland, S. E. Russek, and M. Donahue, "Manipulation of Magnetic Particles by Patterned Arrays of Magnetic Spin-Valve Traps," *J. Magn. Magn. Mater.* (in press).

J. M. Shaw, R. Geiss, and S. E. Russek, "Dynamic Lorentz Microscopy of Micromagnetic Structure in Magnetic Tunnel Junctions" *Appl. Phys. Lett.* **89**, 212503 (November 2006).

P. Voskoboynik, R. D. Joos, W. E. Doherty, and R. B. Goldfarb, "Low Magnetic Moment PIN Diodes for High Field MRI Surface Coils," *Medical Phys.* **33**, 4499-4501 (December 2006).

M. R. Pufall, W. H. Rippard, S. E. Russek, S. Kaka, and J. A. Katine, "Electrical Measurement of Spin-Wave Interactions of Proximate Spin Transfer Nanooscillators," *Phys. Rev. Lett.* **97**, 087206 (August 2006).

A. J. Barker, B. Cage, S. E. Russek, R. Garg, R. Shandas, and C. R. Stoldt, "Tailored Nanoscale Contrast Agents for Magnetic Resonance Imaging," Proc. ASME International Mechanical Engineering Congress and Exposition, Orlando, FL (November 2005).

S. Kaka, M. R. Pufall, W. H. Rippard, T. J. Silva, S. E. Russek, and J. A. Katine, Mutual Phase-Locking of Microwave Spin Torque Nano-Oscillators, *Nature* **437**, 389-392 (September 2005).

B. Cage, S. E. Russek, D. Zipse, J. M. North, and N. S. Dalal, "Resonant Microwave Power Absorption and Relaxation of the Energy Levels of the Molecular Nanomagnet Fe8 Using Superconducting Quantum Interference Device-Based Magnetometry," *Appl. Phys. Lett.* **87**, 082501 (August 2005).

W. E. Bailey, S. E. Russek, X.-G. Zhang, and W. H. Butler, "Experimental Separability of Channeling Giant Magnetoresistance in Co/Cu/Co," *Phys. Rev. B* **72**, 012409 (July 2005).

E. Mirowski, J. Moreland, A. Zhang, S. E. Russek, and M. J. Donahue, "Manipulation and Sorting of Magnetic Particles by a Magnetic Force Microscope on a Microfluidic Magnetic Trap Platform," *Appl. Phys. Lett.* **86**, 243901 (June 2005).

B. Cage, S. E. Russek, D. Zipse, and N. S. Dalal, "Advantages of Superconducting Quantum Interference Device-Detected Magnetic Resonance Over Conventional High-Frequency Electron Paramagnetic Resonance for Characterization of Nanomagnetic Materials," *J. Appl. Phys.* **97**, 10M507 (May 2005).

N. A. Stutzke, S. E. Russek, D. P. Pappas, and M. Tondra, "Low-Frequency Noise Measurements on Commercial Magnetoresistive Magnetic Field Sensors," *J. Appl. Phys.* **97**, 10Q107 (May 2005).

W. L. Johnson, S. A. Kim, S. E. Russek, and P. Kabos, "Brillouin Light Scattering from Pumped Uniform-Precession and Low-k Magnons in $Ni_{81}Fe_{19}$," *Appl. Phys. Lett.* **86**, 102507 (March 2005).

S. E. Russek, S. Kaka, W. H. Rippard, M. R. Pufall, and T. J. Silva, "Finite-Temperature Modeling of Nanoscale Spin-Transfer Oscillators," *Phys. Rev. B* **71**, 104425 (March 2005).

M. R. Pufall, W. H. Rippard, S. Kaka, T. J. Silva, and S. E. Russek, "Frequency Modulation of Spin-Transfer Oscillators," *Appl. Phys. Lett.* **86**, 082506 (February 2005).

D. Min, A. McCallum, S. E. Russek, and J. Moreland, "Micromechanical Torque Magnetometer with Sub-Monolayer Sensitivity," *J. Magn. Magn. Mater.* **286**, 329-335 (February 2005).

S. Kaka, M. R. Pufall, W. H. Rippard, T. J. Silva, S. E. Russek, J. A. Katine, and M. Carey, "Spin Transfer Switching of Spin Valve Nanopillars Using Nanosecond Pulsed Currents," *J. Magn. Magn. Mater.* **286**, 375-380 (February 2005).

B. Cage and S. Russek, "Design for a Multifrequency High Magnetic Field Superconducting Quantum Interference Device-Detected Quantitative Electron Paramagnetic Resonance Probe: Spin-Lattice Relaxation of Cupric Sulfate Pentahydrate ($CuSO_4 \cdot 5H_2O$)," *Rev. Sci. Instrum.* **75**, 4401-4405 (November 2004).

W. H. Rippard, M. R. Pufall, S. Kaka, T. J. Silva, and S. E. Russek, "Current-Driven Microwave Dynamics in Magnetic Point Contacts as a Function of Applied Field Angle," *Phys. Rev. B* **70**, 100406 (September 2004).

A. T. McCallum and S. E. Russek, "In Situ Observation of Nano-Oxide Formation in Magnetic Thin Films," *IEEE Trans. Magn.* **40**, 2239-2241 (July 2004).

M. R. Pufall, W. H. Rippard, S. Kaka, S. E. Russek, T. J. Silva, J. Katine, and M. Carey, "Large-Angle, Gigahertz-Rate Random Telegraph Switching Induced by Spin-Momentum Transfer," *Phys. Rev. B* **69**, 214409 (June 2004).

D. Das, A. Saha, C. M. Srivastava, R. Raj, S. E. Russek, and D. Bahadur, "Magnetic and Electrical Transport Properties of $La_{0.67}Ca_{0.33}MnO_3$ (LCMO): xSiCN Composites," *J. Appl. Phys.* **95**, 7106-7108 (June 2004).

A. T. McCallum and S. E. Russek, "Current Density Distribution in a Spin Valve Determined Through *In Situ* Conductance Measurements," *Appl. Phys. Lett.* **84**, 3340-3342 (April 2004).

E. Mirowski, J. Moreland, S. E. Russek, and M. J. Donahue, "Integrated Microfluidic Isolation Platform for Magnetic Particle Manipulation in Biological Systems," *Appl. Phys. Lett.* **84**, 1786-1788 (March 2004).

W. H. Rippard, M. R. Pufall, S. Kaka, S. E. Russek, and T. J. Silva, "Direct-Current Induced Dynamics in $Co_{90}Fe_{10}/Ni_{80}Fe_{20}$ Point Contacts," *Phys. Rev. Lett.* **92**, 027201 (January 2004).

Microsystems for Bio-Imaging and Metrology

Goals

The Microsystems for Bio-Imaging and Metrology Project designs, fabricates, and tests microelectromechanical systems (**MEMS**) for studying microscopic and nanoscopic magnetic phenomena. Project members are taking an approach based on chip-scale microsystems and nanosystems to advance instrumentation by improving sensitivity, portability, cost, and traceability to fundamental constants. The research has traditionally focused on the data storage, electronics, and communication industries and is currently exploring applications in medicine and bioengineering. Recent programs include magnetic manipulation and measurement of single molecules in microfluidic environments, engineered radio-frequency tags for magnetic resonance imaging (**MRI**) and microfluidics, single molecule enzymology, precision cantilevers for transfer standards and intrinsic force measurements, microfabrication of chip-scale atomic devices, and integration of chip-scale MRI microscopy systems.

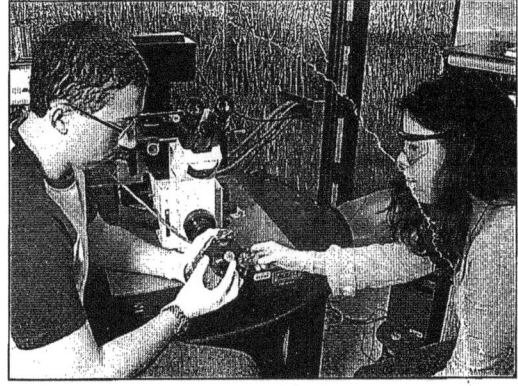

Bio-atomic force microscope for microfluidics experiments on single molecules.

Customer Needs

Methods for measuring the properties and estimating the performance of nanomagnetic materials and devices are in the early development stage. Conventional magnetometers lack the necessary sensitivity for direct measurement of single nanoparticles or nanodevices. We target medical and security applications where we believe a significant impact could be made in the next five years in developing new measurement tools for basic research, rapid detection, assay, and diagnosis; early detection of diseases; and drug development and approval.

We are developing new methods for magnetic measurements at the nanometer scale. Examples include cantilever magnetometers, where a sample is integrated with the sensor for maximum coupling, which hold promise for magnetic measurements at room temperature at the level of 1000 atoms. We develop: (1) methods for understanding the physical processes that give rise to the unique properties of nanomagnetic materials, (2) new instrumentation for nanometer scale MEMS magnetometry, and (3) new protocols for detecting the distinct signatures of magnetic nanoparticles.

The medical and security communities are acutely interested in harnessing the unique properties of nanomaterials, with nanomagnetism being one of the key physical characteristics of interest. We focus on the properties of nanoparticles dispersed in liquids or solids at the bulk and single-particle limits, with the ultimate goal of developing them for new imaging modalities. These include magnetic particle imaging (MPI) and MRI for medical applications, microfluidic devices that use magnetic particles for single cell and single molecule bioassay applications, radio-frequency tags for cell tracking in research animal models, and forensic identification of controlled substances.

Technical Strategy

The project's technical strategy is to develop: (1) new, ultrasensitive magnetometers based on MEMS chips, MPI, and MRI, (2) integrated microsystems and new imaging modalities based on microscale and nanoscale magnetic particles, (3) quantitative cantilevers and calibration methods for quantifying the performance of atomic force microscopes, (4) magnetic microfluidic platforms for single-molecule studies, and (5) integrated microsystems based on atomic transitions.

MEMS Magnetometry of Patterned Sub-Micrometer Dots

Sub-micrometer scale magnetic measurements have proven to be a challenge for conventional magnetometers, and new methods are being employed to probe magnetism on this scale. We will provide new magnetometers based on highly specialized MEMS chips fabricated at NIST. Such instruments will be inexpensive, since MEMS can be batch-fabricated in large quantities. In addition, large-scale magnetic wafer properties can be transferred to smaller MEMS magnetometers so that

Technical Contact:
John Moreland

Staff-Years (FY 2006):
1.0 professional
0.5 technician
2.0 research associates
1.0 graduate student

nanometer-scale measurements can be calibrated with reference to fundamental units. In particular, the focus is on developing torque and force magnetometers, magnetic-resonance spectrometers, and MRI microscopes on MEMS chips. Over the long term, we expect that this technology will lead to atomic-scale magnetic instrumentation for the measurement and visualization of fundamental magnetic phenomena.

MEMS Based Magnetic Moment Standard Reference Materials

We are developing a method for defining the magnetic moment of a reference material based on a torsional resonator with a patterned magnetic film on its surface. Given accurate measurements of the magnitude of the applied field, the moment of inertia of the resonator, and the magnetic stiffening effect of the film on the resonance frequency of the resonator, an absolute measurement of the anisotropy energy and magnetic moment of the film can be made independently. With this approach, the anisotropy energy and the magnetic moment of the film are measured directly, eliminating the need for a detailed knowledge of the film's saturation magnetization or sample volume. Calibrated moments as small as 10^{-6} ampere meter squared can be measured. In principle, all of the measurements needed for the reference material are traceable to an atomic clock frequency reference.

Microfabricated Radio Frequency Tags for MRI Tracking of Single Cells

We are developing a new biological detection methods that would have application in two areas: (1) to replace conventional serial fluorescence-based (and also more recently suggested optically probed micro-metallic barcode) techniques with a parallel detection scheme capable of simultaneously detecting multiple bio-agents without the need for optical access to the system, and (2) to improve the specificity of cell labeling strategies for MRI. Custom designed magnetic micro-particle tags that induce distinct magnetic resonance frequency shifts in the nuclear magnetic resonance (**NMR**) of protons in water surrounding the labeled biological agent are being microfabricated and tested. Magnetic detection would be based on the tag particle set being specifically tailored (in size, shape, and composition using microlithography) to give distinct NMR frequency shifts and resonance line shapes. Such a magnetic resonance system would enable noninvasive and inherently parallel detection both *in vitro* and *in vivo*.

Integrated Chip-Scale MRI Microscopes

We are developing methods for integration of DC and radio-frequency magnetic field sources and MEMS cantilever oscillators to perform magnetic resonance imaging and spectroscopy of small samples attached to the oscillators. The main goal is to push the technology to measure magnetic phenomena at the nanometer scale. Cantilevers operated at resonance have the potential for detecting single magnetic nanoparticle tags with 10^5 spins at room temperature, and there are prospects for single spin detection at low temperatures. Shrinking the magnetic subsystems of a magnetic resonance instrument into a chip-scale microsystem provides benefits in terms of reduced size, batch fabrication cost, and better performance via an increased detector bandwidth and sensitivity.

Calibrated Cantilevers for Force Transfer Standards at the Nanonewton Range

The ability to demonstrate traceable quantitative measurement of forces in the regime below 10 millinewtons is a tremendous challenge for a variety of manufacturers. We are developing metrology for use by primary, secondary, and industrial laboratories to support quantitative measurement of these forces. The goal is an instrument that incorporates NIST's force and length metrology to traceably determine the spring constants of silicon reference cantilevers and torsion oscillators. Silicon reference cantilevers have been proposed by the Materials Science and Engineering Laboratory as Standard Reference Materials (**SRMs**) for calibrated adhesion tests. Calibrated piezoresistive cantilever load cells could ultimately serve as a standard reference material for the dissemination of force to atomic force microscopes (**AFMs**) and instrumented indentation machines. Torsion oscillators may be used as SRMs for magnetometers.

Cantilever Sensors for Intrinsic Force Measurements at the Femtonewton Range of Single Molecules

Among the experiments made possible by the revolution in scanned probe microscopy, measurements of bond rupture (forces between atoms), binding rupture (forces between molecules), and molecular conformation changes (force induced structural changes of individual molecules) are perhaps the most remarkable. Such forces might then serve as intrinsic standards, effectively rendering this class of instruments self-calibrating. We are developing the needed femtonewton force sensors for the various applications. For instance, an ultrathin cantilever with an integral interferometric displacement sensor is appropriate for use in the binding rupture experiments since the unit must be compatible with aqueous biological environments. For the metal bond rupture experiments, a feedback-controlled torsion balance using either electromagnetic or electrostatic forces to provide the null compensation is more appropriate. These sensors will be calibrated in an ultrahigh vacuum, atomic force test and calibration apparatus in the Manufacturing Engineering Laboratory and the single molecule laser tweezers apparatus at JILA (jointly operated by the University of Colorado and NIST).

MAGNETIC TEMPLATES FOR NANOMETER SCALE MANIPULATION AND ASSEMBLY OF MAGNETIC PARTICLES AND DEVICES

We are developing a nanometer scale assembly platform based on an array of "switchable" magnetic dots for manipulating magnetic components. In the current configuration, a magnetic force microscope (**MFM**) tip is used to distribute magnetic polystyrene beads in an array of patterned Permalloy (Ni-Fe) traps on a thin membrane as part of a microfluidic flow cell. We have demonstrated the ability to translate particles with nanometer precision and sort micrometer sized magnetic particles based on size differences in the array with the MFM. In addition, we are currently developing fluidic cells with the potential for large-scale integration based on "spin-valve" technology using magnetically balanced spin-valve structures that act as switchable permanent magnets with a ferromagnetic "on" or an antiferromagnetic "off" net magnetization state so that magnetic particles in the fluid cell can be electronically confined and released for transporting, sorting, or assembly applications. Spin-valve magnetic traps and a matrix addressable architecture similar to MRAM would then eliminate the need for a scanning MFM tip for translating particles. In principle, these techniques would be the basis for nanometer scale robotic assembly of components to achieve novel biological, chemical, electrical, or mechanical functionality at the single molecule level.

SINGLE MOLECULE ENZYMOLOGY: MECHANICS OF REPLICATION

The structure of DNA and individual proteins is being examined in increasing detail; however our knowledge of the mechanics of their interactions is often limited to bulk experiments. We are developing a new single-molecule assay to assist in elucidating the mechanics of enzyme activity. The flexibility of this *in-vitro* assay will be used to determine the operation of individual enzymes, complementing existing genetic techniques. The ultimate goal of this project is to work towards a more complete understanding of the mechanics of DNA polymerization and how errors in this process can lead to a range of neurological diseases, eventually developing an *in-vivo* assay. To control and manipulate the DNA substrate we will be working with superparamagnetic beads in a microfluidic environment. By attaching a magnetic bead to one end and applying a magnetic field we can stretch out the DNA. This provides a static one-dimensional template along which enzyme activity can be easily observed for longer periods and correlated with the underlying base-pair sequence.

MEMS DESIGN, FABRICATION, AND PACKAGING FOR CHIP-SCALE ATOMIC CLOCKS AND MAGNETOMETERS

Many important electronic devices (such as global positioning receivers, wireless receivers, portable magnetometers, and compact gyroscopes and accelerometers) would greatly improve if very small, highly accurate, low-power, and low-cost measurement references were available. These devices are typically very large (often laboratory-scale), often consume kilowatts of power or require cryogenic cooling, and are generally too expensive for wide-scale applications. The challenge is to shrink the size of the sensors and standards based on atomic properties from the laboratory scale (10 cubic meters) to the 1 cubic centimeter scale of a computer chip. While the chip-scale atomic clock (as an early example of a chip-scale atomic device) is only the size of a rice grain, it is a complex structure comprised of more than a dozen components; more advanced chip-scale atomic devices will likely be more complex. We are developing new methods for designing, fabricating, and assembling chip-scale atomic devices optimized for different applications. A key part of the chip-scale atomic device is the ultra-miniature atomic vapor cell containing the active sensing atoms. We are developing new technologies to design, fabricate, and assemble these crucial vapor cells into various chip-scale atomic device packages and applications.

ACCOMPLISHMENTS

■ **Novel Fabrication of Micromechanical Oscillators with Nanoscale Sensitivity at Room Temperature** — Over the past decade, several experimental methods have been developed to probe material properties on the micrometer and nanometer scales. Many of these novel methods employ the use of micromechanical cantilevers to achieve the desired sensitivity. Because these experiments are limited by the thermal noise of the cantilever itself, low temperatures often must be used, making the results less relevant to industrial applications. Thus, there is a strong demand for microcantilevers that are sensitive enough to obtain useful results at room temperature. A further complication arises when the experiments require that a micrometer-sized magnetic material be placed onto the cantilever. Doing this on an individual basis is not only time-consuming, but it also jeopardizes the uniformity and consistency of the results. Therefore, there is a clear need to develop a process to batch-fabricate ultrasensitive cantilevers with magnetic dots prealigned and deposited as part of the microfabrication process.

We have developed a process to meet these demands. Careful consideration was given to the design of the oscillator shape, and finite-element modeling was used to study the resonant shapes and to make sure that the resonance frequencies were in the desired ranges for the specific applications. Our ultrathin (less than 1 micrometer) single-crystal silicon cantilevers with integrated magnetic structures are the first of their kind. They were fabricated with a novel high-yield process in which magnetic film patterning and deposition are combined with cantilever fabrication. These novel devices have been developed for use as cantilever magnetometers and as force sensors in nuclear magnetic resonance force microscopy. These two applications have achieved nanometer-scale resolution with these cantilevers. Current magnetic moment sensitivity achieved for the devices, when used as magnetometers, is 10^{-15} ampere meter squared at room temperature, which is more than a thousand-fold improvement in sensitivity compared to conventional magnetometers. Finite element modeling was used to improve design parameters, ensure that the devices meet experimental demands, and correlate mode shape with observed results. The photolithographic fabrication process was optimized, yielding an average of 85 percent and alignment better than 1 micrometer. Post-fabrication focused ion-beam milling was used to further pattern the integrated magnetic structures when nanometer scale dimensions were required.

■ **MEMS *In-Situ* Magnetometers to Measure Interlayer Coupling in Fe/Cr/Fe Trilayers** — The characterization of thin magnetic films, patterned recording media, and nanometer-scale magnetic devices has proven to be a challenge for conventional magnetometers at nanoscale and atomic dimensions. The limitation on the sensitivity of magnetic moment sensors used in these instruments is fundamentally understood by comparing the energy necessary to excite the sensor relative to the energy necessary to excite the specimen for measurement purposes. Conventional magnetometers are designed for measurements of large specimens and therefore have a relatively low signal-to-noise ratio for small specimens. Sensitivity can be improved tremendously by integrating the specimens with the measurement sensor using microfabrication methods. Further, *ex-situ* measurements outside of the growth chamber require sample transfer, so there is the potential for substantial oxidation of thin films and devices.

We have tailored an ultrasensitive magnetometer for the *in-situ* study of thin-film interface magnetism and interlayer magnetic exchange coupling. The magnetometer is based on a customized micro-resonator made with silicon MEMS fabrication techniques. The measured changes in magnetic moments using the MEMS magnetometer were compared to theoretical calculations and measurements made by more conventional methods. In

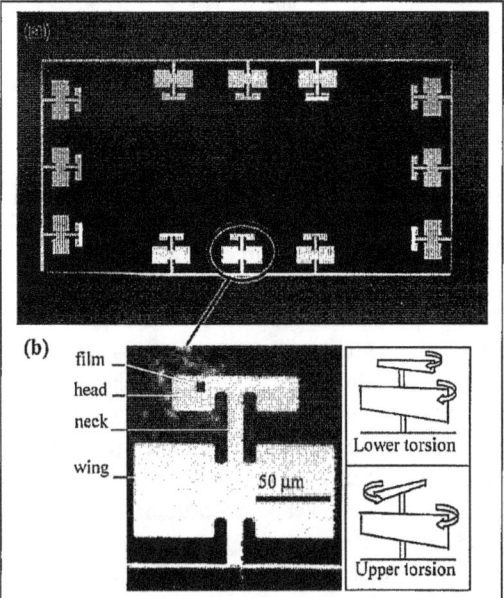

Scanning electron micrographs of the double torsional oscillator geometry. (a) A typical chip containing 12 devices, all with double-side access. (b) A closer view of a double torsional oscillator with a 5 micrometer by 5 micrometer by 30 nanometer film on the head. The illustration on the right side indicates the shape of the two main torsional modes of operation.

particular, we followed a previous formalism that includes interface roughness (interdiffusion length of different atoms) for predicting the oscillation periods of interlayer magnetic exchange coupling of Fe/Cr/Fe thin film multilayers as a function of Cr spacer thickness. Our results indicate that the magnetometer is well suited for studying the subtle changes in the magnetic moment of multilayer films during film growth. Specifically we determined the short and long period oscillation wavelength of the interlayer exchange coupling in the Cr layer to be 2.1 monolayers and 13.8 monolayers, respectively. These results illustrate the utility of MEMS magnetometers for studying ultrathin magnetic films and multilayer devices.

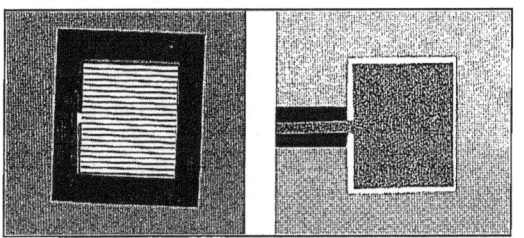

Custom cantilever assembly for submonolayer in-situ magnetometery of multilayer films (front and back views). The paddle is 1 millimeter on a side.

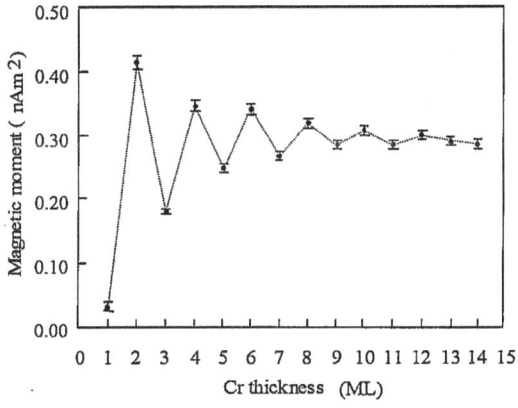

Graph of experimental magnetic interlayer exchange coupling measured by the micro-resonator magnetometer as a function of Cr layer thickness.

■ **Manipulation of Magnetic Particles by Patterned Arrays of Magnetic Spin-Valve Traps** — To address some of the limitations of current single-molecule tweezers techniques, we developed a novel magnetic tweezers based on a chip-scale microfluidic platform that can trap, measure, manipulate and sort magnetic particles in an array. The platform consists of an array of magnetic spin-valve elements separated from the biological sample by an optically transparent thin membrane, effectively isolating the electronic or magnetic components from the fluid bead solution. The particles are confined by local magnetic field gradients in an array of magnetic spin-valve structures. The spin valves have a bistable ferromagnetic "on" and antiferromagnetic "off" net magnetization states in the absence of an externally applied magnetic field, which allows magnetic particles to be selectively confined or released for transport or sorting.

Single particles can be rotated by applying a global magnetic field, and thus, the platform may be potentially used as a magnetic molecular tweezer where particles are rotated about the x, y or z axes while attached to multiple points of a biological molecule such as the ends of a strand of DNA to impart torsional forces. Alternatively, spin-valve trap arrays can be adapted to a low power MRAM switching architecture for massively parallel particle sorting applications. Previous work with spin-valve trapping elements in terms of biological microfluidic applications focuses on their ability to detect the presence of magnetic particles as they attach to locations with specific biological antibodies. In contrast, the current platform incorporates spin-valve elements that can be switched between bistable states to provide a local magnetic field gradient sufficiently large to trap a magnetic particle that can be used not only for the purpose of investigating conformational dynamics of single biological molecules but also to capture and sort biological molecules for gene sequencing or bioassay applications.

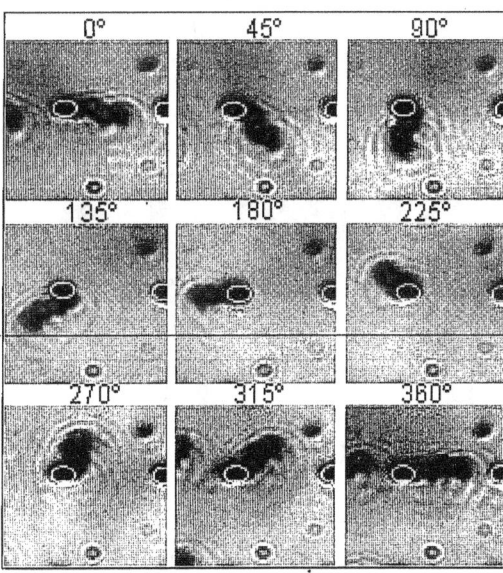

Video micrographs showing the rotation of a strand of magnetic particles while trapped at the edge of a single "on" state spin-valve element (highlighted by the white ellipses). The magnetic beads are 2.8 micrometers in diameter.

■ **Wafer-Level Fabrication and Filling of Cesium Vapor Cells for Chip-Scale Atomic Devices** — The introduction of atomic vapor systems to MEMS has enabled a new class of hybrid microchips referred to as chip-scale atomic devices (**CSADs**) with integrated optical, radio-frequency, electronic, and quantum based atomic components. At the heart of a CSAD is a miniature cell containing alkali vapor. Suitable techniques for fabricating vapor cells must be developed such that they will not only be in a physical form suitable for integration into the rest of a MEMS assembly, but also in a manner that is amenable to true wafer-scale mass production. Whereas conventional cells are made by glass blowing, millimeter scale vapor cells are more easily fabricated by means of standard photolithography, silicon etching, and silicon-glass bonding techniques in a cleanroom. However, filling the cells with the highly volatile alkali metals and buffer gasses without exposure to air, and subsequent hermetic sealing of the cells, are formidable MEMS fabrication problems. The high temperature required for anodic bonding of silicon to glass is at odds with the low melting points of Cs and Rb. Thus, to date, cell fabrication requires technically complex methods that are limited to sequential rather than parallel processes to introduce the alkali metal into the cell and to subsequently seal the cell hermetically.

To overcome these disadvantages and simplify the cell-filling process, we have developed and tested a cesium batch-filling technique based on thin-film deposition and subsequent decomposition of cesium azide (CsN_3). CsN_3 is a solid at room temperature and is stable in air. It decomposes to produce pure cesium and nitrogen gas when heated to 450 degrees Celsius or through ultraviolet photolysis at room temperature. The azide method of cell filling eliminates the need for pipetting or microfluidic manipulation of liquid alkali metals since the Cs is liberated *in situ* after the cell has been hermetically sealed, thus simplifying the filling process and dramatically reducing the tooling cost.

Furthermore, the azide method, being based on thin-film deposition, is inherently wafer-level. Due to the lack of thermophysical and thermochemical data on CsN_3, as well as the relative lack of published literature concerning CsN_3 material properties, much of this work was exploratory in nature. While the photolysis of other metal azides has been studied in detail, the photolysis of CsN_3 has not been previously demonstrated or studied before and as such the mechanisms for the photodissociation are just now coming to light.

■ **Chip-Scale Atomic Magnetometer** — In collaboration with the Physics Laboratory, we have constructed, using MEMS techniques, a small low-power magnetic sensor based on alkali atoms. We used a coherent population trapping resonance to probe the interaction of the atoms' magnetic moment with a magnetic field, and we detected changes in the magnetic flux density with a sensitivity of 50 picotesla per root hertz at 10 hertz. The magnetic sensor has a size of 12 cubic millimeters and dissipates 195 milliwatts of power. Further improvements in size, power dissipation, and magnetic field sensitivity are immediately foreseeable; such a device could provide a handheld battery-operated magnetometer with an atom shot-noise-limited sensitivity of 0.05 picotesla per root hertz.

Photograph of a Cs MEMS cell as viewed through the glass window showing in-situ production of Cs.

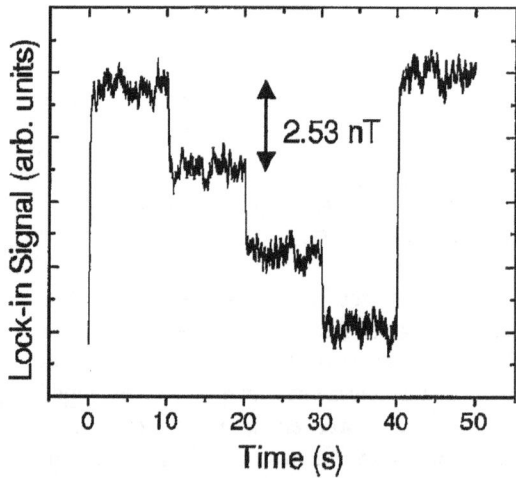

Lock-in amplifier signal plotted as a function of time as magnetic flux is stepped in 10 second. The magnetic flux density during the measurement is nominally 73.9 millitesla.

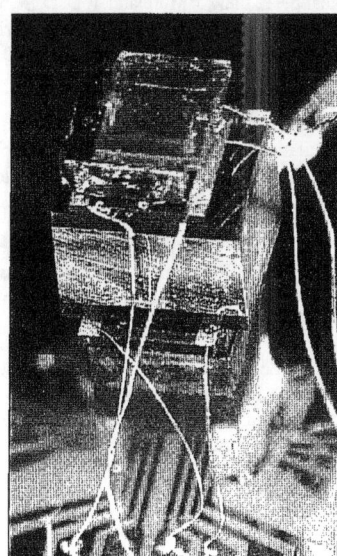

Photograph of the magnetometer physics package, which is the size of a grain of rice.

■ **Force Standard Reference Material for AFM Calibrations** — We are designing a piezoresistive cantilever force sensor that can serve as a force and/or stiffness SRM. The beam thickness in the gauge area was reduced to produce high sensitivity. The trade-off for this sensitivity is a more complex artifact with the need to maintain additional critical dimensions. The numbered fiducial marks shown in the figure indicate points along the cantilever's length where a calibrated stiffness will be determined. The functional dependence of stiffness on test location is more complicated in this design, but can be modeled assuming the gauge area acts as a cantilever hinge and the remainder of the beam is a rigid lever arm. The exact functional relationship will be examined using the electronic force balance in the Manufacturing Engineering Laboratory. As with the previously described stiffness SRMs, the goal is to settle on a fabrication strategy that yields a high degree of uniformity within a wafer. We hope to use resonance frequency, this time conveniently measured from the piezoresistor, with a bridge as a process monitor to check for mechanical uniformity within a wafer run. This sensor is currently in prototype production and testing results are not yet available.

AWARD

U.S. Department of Commerce Silver Medal for the design, construction, and testing of the first operational chip-scale atomic clock and chip-scale magnetometer based on atomic transitions, 2005 (John Moreland, Li-Anne Liew, EEEL; John Kitching, Peter Schwindt, Hugh Robinson, and Leo Hollberg, Physics Laboratory).

RECENT PUBLICATIONS

S. Knappe, P. Schwindt, V. Gerginov, V. Shah, A. Brannon, B. Linseth, L. Liew, H. Robinson, J. Moreland, Z. Popovic, L. Hollberg, and J. Kitching, "Chip-Scale Atomic Devices," *Proc. International Symposium on Quantum Electronics*, 2006 (in press).

S. Knappe, P. D. D. Schwindt, V. Gerginov, V. Shah, L. Liew, J. Moreland, H. G. Robinson, L. Hollberg, and J. Kitching, "Microfabricated Atomic Clocks and Magnetometers," *Proc. IEEE/LEOS Optical MEMS Conf.*, Big Sky, MT, August 2006 (in press).

E. Mirowski, J. Moreland, S. E. Russek, and M. Donahue, "Manipulation of Magnetic Particles by Patterned Arrays of Magnetic Spin-Valve Traps," *J. Magn. Magn. Mater.* (in press).

S. Lee, T. M. Wallis, J. Moreland, P. Kabos, and Y. C. Lee, "Asymmetric Dielectric Trilayer Cantilever Probe for Calorimetric High Frequency Field Imaging," *IEEE J. Microelectromech. Syst.* (in press).

Y.-J. Wang, M. Eardley, S. Knappe, J. Moreland, L. Hollberg, and J. Kitching, "Magnetic Resonance in an Atomic Vapor Excited by a Mechanical Resonator," *Phys. Rev. Lett.* **97**, 227602 (December 2006).

L. Yuan, L. Gao, R. Sabirianov, S. Liou, M. Chabot, D. Min, J. Moreland, and B. S. Han, "Microcantilever Torque Magnetometry Study of Patterned Magnetic Films," *IEEE Trans. Magn.* **42**, 3234-3236 (October 2006).

T. M. Wallis, J. Moreland, and P. Kabos, "Einstein–de Haas Effect in a NiFe Film Deposited on a Microcantilever," *Appl. Phys. Lett.* **89**, 122502 (September 2006).

L. Liew, J. Moreland, and V. Gerginov, "Wafer-Level Fabrication and Filling of Cesium-Vapor Resonance Cells for Chip-Scale Atomic Devices," *Proc. 20th Eurosensors Conf.*, Göteborg, Sweden, paper W1B-O2 (September 2006).

S. Knappe, P. D. D. Schwindt, V. Gerginov, V. Shah, L. Liew, J. Moreland, H. G. Robinson, L. Hollberg, and J. Kitching, "Microfabricated Atomic Clocks and Magnetometers," *J. Optics A: Pure and Applied Optics* **8**, S318-S332 (July 2006).

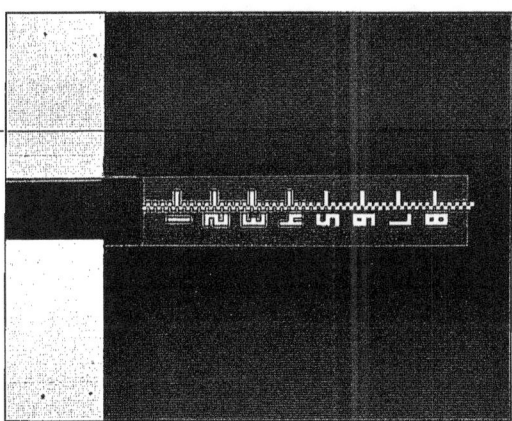

Prototype piezoresistive cantilever SRM for calibrating AFMs. The cantilever is 500 micrometers long.

J. R. Pratt, J. A. Kramar, G. Shaw, R. Gates, P. Rice, and J. Moreland, "New Reference Standards and Artifacts for Nanoscale Property Characterization," Tech. Proc. 9th NSTI Nanotechnology Conf. and Trade Show, May 2006, vol. 1, pp. 764-767 (July 2006).

J. Kitching, S. Knappe, P. D. D. Schwindt, Y.-J. Wang, H. Robinson, L. Hollberg, L. Liew, J. Moreland, A. Brannon, J. Breitbarth, B. Lindseth, Z. Popovic, V. Shah, V. Gerginov, and M. Eardley, "Chip-Scale Atomic Devices," Proc. Solid State Sensors, Actuators and Microsystems Workshop, Hilton Head Island, SC, pp. 108-113 (June 2006).

S. Lee, Y. C. Lee, T. M. Wallis, J. Moreland, and P. Kabos, "Near-Field Imaging of High-Frequency Magnetic Fields with Calorimetric Cantilever Probes," J. Appl. Phys. **99**, 08H306 (April 2006).

J. Moreland, J. Kitching, P. D. D. Schwindt, S. Knappe, L. Liew, V. Shah, V. Gerginov, Y. Wang, and L. W. Hollberg, "Chip Scale Atomic Magnetometers," Military Sensing Symposia Specialty Group on Battlefield Acoustic and Seismic Sensing, Magnetic and Electric Field Sensors, August 2005, Laurel, MD, pp. 1-10 (April 2006).

J. Kitching, S. Knappe, L. Liew, J. Moreland, H. G. Robinson, P. D. D. Schwindt, V. Shah, V. Gerginov, and L. W. Hollberg, "Chip-Scale Atomic Frequency References: Fabrication and Performance," Proc. 19th European Frequency and Time Forum, Besancon, France, pp. 575-580 (December 2005).

M. Chabot, J. Moreland, L. Gao, S. H. Liou, and C. W. Miller, "Novel Fabrication of Micromechanical Oscillators with Nanoscale Sensitivity at Room Temperature," J. Microelectromech. Syst. **14**, 1118 (October 2005).

J. Kitching, S. Knappe, L.-A. Liew, P. D. D. Schwindt, V. Gerginov, V. Shah, J. Moreland, A. Brannon, J. Breitbarth, Z. Popovic, and L. Hollberg, "Chip-Scale Atomic Frequency References," Proc. Institute of Navigation (ION-GNSS), Long Beach, CA, pp. 1662-1669 (September 2005).

V. Gerginov, S. Knappe, P. D. D. Schwindt, V. Shah, L. Liew, J. Moreland, H. G. Robinson, L. Hollberg, J. Kitching, A. Brannon, J. Breitbarth, and Z. Popovic, "Component-Level Demonstration of a Microfabricated Atomic Frequency Reference," Proc. IEEE International Frequency Control Symposium Vancouver, BC, pp. 758-766 (August 2005).

J. Kitching, S. Knappe, L. Liew, P. D. D. Schwindt, V. Shah, J. Moreland, and L. W. Hollberg, "Microfabricated Atomic Clocks," 18th International Conference on Micro-Electro-Mechanical Systems, January 2005, Miami, FL, pp. 1-7 (July 2005).

J. Kitching, S. Knappe, L. Liew, J. Moreland, P. D. D. Schwindt, V. Shah, V. Gerginov, and L. Hollberg, "Microfabricated Atomic Frequency References," Metrologia **42**, S100-S104 (June 2005).

E. Mirowski, J. Moreland, A. Zhang, S. E. Russek, and M. J. Donahue, "Manipulation and Sorting of Magnetic Particles by a Magnetic Force Microscope on a Microfluidic Magnetic Trap Platform," Appl. Phys. Lett. **86**, 243901 (June 2005).

D. Min and J. Moreland, "Quantitative Measurement of Magnetic Moments with a Torsional Resonator: Proposal for an Ultralow Moment Reference Material," J. Appl. Phys. **97**, 10R504 (May 2005).

S. Knappe, P. D. D. Schwindt, V. Shah, L. W. Hollberg, J. Kitching, L. Liew, and J. Moreland, "A Chipscale Atomic Clock Based on 87Rb with Improved Frequency Stability," Optics Express **13**, 1249-1253 (February 2005).

T. M. Wallis, J. Moreland, B. Riddle, and P. Kabos, "Microwave Power Imaging with Ferromagnetic Calorimeter Probes on Bimaterial Cantilevers," J. Magn. Magn. Mater. **286**, 320-323 (February 2005).

D. Min, A. McCallum, S. E. Russek, and J. Moreland, "Micromechanical Torque Magnetometer with Sub-Monolayer Sensitivity," J. Magn. Magn. Mater. **286**, 329-335 (February 2005).

P. D. D. Schwindt, S. Knappe, V. Shah, L. W. Hollberg, J. Kitching, J. Moreland, and L. Liew, "Chipscale Atomic Magnetometer," Appl. Phys. Lett. **85**, 6409-6411 (December 2004).

S. Knappe, V. Shah, P. D. D. Schwindt, L. Hollberg, J. Kitching, L.-A. Liew, and J. Moreland, "A Microfabricated Atomic Clock," Appl. Phys. Lett. **85**, 1460-1462 (August 2004).

L. Gao, D. Q. Feng, L. Yuan, T. Yokota, R. Sabirianov, S. H. Liou, M. D. Chabot, D. Porpora, and J. Moreland, "A Study of Magnetic Interactions of Ni80Fe20 Arrays Using Ultrasensitive Microcantilever Torque Magnetometry," J. Appl. Phys. **95**, 7010-7012 (June 2004).

L.-A. Liew, S. Knappe, J. Moreland, H. Robinson, L. Hollberg, and J. Kitching, "Microfabricated Alkali Atom Vapor Cells," Appl. Phys. Lett. **84**, 2694-2696 (April 2004).

E. Mirowski, J. Moreland, S. E. Russek, and M. J. Donahue, "Integrated Microfluidic Isolation Platform for Magnetic Particle Manipulation in Biological Systems," Appl. Phys. Lett. **84**, 1786-1788 (March 2004).

Appendix A:
Calibration Services

The Electromagnetics Division provides a number of for-fee calibration services for radio-frequency and microwave standards. Below is a listing of those services conducted by the division, with the technical contacts. More information can be found in the NIST Calibration Services User's Guide SP250, available from the Calibration Program at NIST, 301-975-2002, calibrations@nist.gov, http://www.ts.nist.gov. The Web site provides a fee schedule.

Thermistor Detectors
Ron Ginley   303-497-3634

Scattering Parameters of Passive Multi-Port Devices
Ron Ginley   303-497-3634

Thermal Noise Measurements
Dave Walker   303-497-5490

Dielectric and Magnetic Material Measurements
Jim Baker-Jarvis   303-497-5621

Microwave Antenna Parameter Measurements
Katie MacReynolds   303-497-3471

Field Strength Parameter Measurements
Dennis Camell   303-497-3214

Information about the availability and shipping requirements for Electromagnetics Division services may be obtained from Puanani DeLara, 303-497-5284.

Appendix B:
Postdoctoral Research Associateships

NIST offers postdoctoral research associateships in collaboration with the National Research Council (NRC). Research topics and associated advisors for the Electromagnetics Division are listed below. Complete information and applications forms for all NIST NRC postdoctoral offerings are available on the Web at http://www7.nationalacademies.org/rap/. Contact a prospective advisor to discuss details of proposed work and the application process. If you do not find a topic that exactly matches your interest, please contact an advisor in a similar discipline. U.S. citizenship is required for NRC postdoctoral appointments.

NIST's Boulder, Colorado, laboratories are located adjacent to the eastern foothills of the Rocky Mountains.

Electromagnetic Properties of Materials

J. R. Baker-Jarvis

Research opportunities exist for theoretical and experimental investigations related to the precise characterization of the macroscopic and microscopic electromagnetic properties of artificial, substrate, liquid, and thin-film materials in the radio-frequency to terahertz frequencies. Suggested research areas include the measurement of ultra low-loss dielectric and magnetic materials, thin films, nanowires, ferroelectric materials, metamaterials, and nanoscale materials. Current interests also include the relationship of quantum-coherence to dielectric loss, the relationships of complex permittivity to material mechanical stress, linear and nonlinear response theory, dielectric measurements of liquids, ferroic materials, and biased ferrite and ferroelectric measurements. In addition to excellent computational resources, a wide range of experimental resonators, well characterized materials, laboratory instrumentation, environmental facilities (*e.g.*, a cryostat, a high-temperature chamber, and a magnetic bias capability), and a diversity of related intellectual resources are available to facilitate research.

Broadband Impedance Measurements of Thin-Film Electronic Materials and Devices

J. C. Booth

Electrical impedance measurements provide important information about the response of materials to electromagnetic stimuli. Therefore, impedance is a key parameter for the evaluation of new thin-film materials that are being developed for next-generation electronic applications. A detailed understanding of the evolution of the electrical impedance of thin-film materials and devices with increasing frequency is required for determining the suitability of new materials for high-speed applications, and can also provide valuable insight into the underlying physical processes relevant for advanced electronic materials. We are developing new broadband impedance measurements of thin-film materials and devices in order to address these issues. Use of on-wafer measurements of thin-film devices allows us to measure the impedance of planar thin-film-based structures at frequencies up to 100 gigahertz. When combined with more traditional impedance analysis at lower frequencies, we can obtain frequency-dependent impedance data over the extremely broad frequency range from several 100 hertz to 100 gigahertz. The electrical impedance of planar measurement structures can be used to extract intrinsic material quantities, such as the conductivity or dielectric permittivity, of the constituent thin-film materials. Such measurements techniques are applicable to a wide range of electronic thin-film materials, and are currently being used to study thin-film material systems such as dielectrics, ferroelectrics and high-temperature superconductors.

MICROWAVE BIOELECTRONICS

J. C. Booth

Electromagnetic characterization can be a viable alternative to chemical or optical detection schemes to analyze biological samples. An exciting new option in dielectric characterization is dielectric spectroscopy: the measurement of a sample's dielectric response across a very broad range of frequencies. This technique has the potential to rapidly characterize a wide variety of biological materials, from macromolecules such as hemoglobin to live bacterial samples to proteins and DNA. By combining lithographically defined broadband microwave structures with microfluidic chambers produced by state-of-the-art micromachining techniques, it is possible to determine the dielectric response of extremely small sample volumes (picoliters) over 9 decades in frequency (100 hertz to 100 gigahertz). Integration with other micro-electromechanical systems (MEMS), such as microfluidic pumps and circuits, could enable rapid analysis of large numbers of different samples in high-throughput screening systems. Knowledge of the dielectric properties of biological samples (*e.g.*, tissue, blood) is also necessary for the development of new microwave or millimeter-wave based medical treatments and/or diagnostics, and could also be applied to determine the effect of electromagnetic signals at the cellular level. For example, such measurements could provide valuable data on the possible health hazards posed by the proliferation of commercial wireless devices. Currently, data on the dielectric properties of many biological substances, if existent, are rarely available at frequencies above 20 gigahertz. By extending the frequency range for dielectric analysis up to at least 110 gigahertz, we will create a new tool for characterizing and understanding the interaction between organic materials and microwave radiation.

SUPERCONDUCTOR CRITICAL CURRENT DENSITY DETERMINATION BY NONLINEAR MICROWAVE RESPONSE

J. C. Booth

One of the fundamental properties of superconductors is the critical current density (J_c), which sets the upper bound for the current-carrying capacity of a superconductor. Our recent measurements of the nonlinear response of YBCO thin films have demonstrated good agreement with theoretical predictions for nonlinear response due to pair-breaking in *d*-wave superconductors, and our measurement system yields values for the pair-breaking current density that agree remarkably well with theoretical predictions of this quantity for YBCO thin films. Our nonlinear measurement technique could provide an entirely new path to determining critical current densities in superconductors, and will likely yield intrinsic J_c values closer to theoretical predictions than conventional J_c measurement techniques. We plan to apply this measurement technique to experimentally determine the critical current density of YBCO, MgB_2, Nb, and other technologically important superconductor thin-film materials, and also to explore the role of the symmetry of the superconducting energy gap on the nonlinear microwave response. In addition, we will also explore the use of these reproducible nonlinear effects in superconductors to develop standard devices to aid in the calibration of next-generation nonlinear measurement systems.

MILLIMETER-WAVE AND MICROWAVE POWER MEASUREMENTS

T. P. Crowley

Research will be performed to develop power measurements at millimeter (mm) wave frequencies above 110 gigahertz. NIST currently operates microwave power standards at frequencies from 50 megahertz to 96 gigahertz, with immediate plans to extend operation to 110 gigahertz. At this time, there is a very large gap in frequency coverage between 110 gigahertz and infrared laser measurements. This region in frequency space has an increasing number of applications in astronomy, biology, and homeland security that will be enhanced by better power measurements. The exact approach we take to develop these new power measurements will depend on developments in available equipment and the abilities of the applicant. However, we anticipate that the new sensors will compare heat produced by millimeter-wave power absorption and DC power absorption. One option is the development of a thin-film bolometer resistor. The bolometer detects power through its change in resistance. The bolometer resistor will be thermally isolated from the bulk of its substrate through micromachining techniques. An alternative approach will be to develop a calorimeter that directly senses the temperature rise due to millimeter-wave absorption. A wide range of microwave and millimeter-wave components, sources, and existing standards are available as well as NIST's in-house capabilities for making integrated circuits for research.

QUANTUM-BASED MICROWAVE AMPLITUDE MEASUREMENTS

T. P. Crowley

This research opportunity will develop a technique for measuring microwave magnetic fields based on quantum-mechanical principles. We have already performed a proof-of-concept experiment using a set of laser-cooled cesium atoms in an atomic clock fountain apparatus. Rabi oscillations between two hyperfine levels of the cesium ground state were measured to determine the RF magnetic field strength in a resonant cavity. Our next step in this research is to improve the measurement accuracy by creating an apparatus specifically designed for making measurements in a waveguide. Either laser-cooled atoms or a gas cell of cesium may be used. A DC magnetic field will be employed to vary the RF measurement frequency. The goal of the new experiment will be to perform a comparison with traditional measurements at accuracy comparable to the current state-of-the-art. Ultimately, we hope to develop a technique that is applicable over a broad frequency range and can be used to characterize transfer standards. The research is a fundamental change in the approach to measuring microwave signal levels. Traditional techniques rely on the heating induced when microwave power is absorbed and are based on comparisons with DC power measurements. Uncertainties are introduced because RF and DC power are not typically dissipated in the same location. We hope that a quantum-based measurement will greatly improve the accuracy of our primary standards and lead to new applications and techniques. Available resources will include a laser system to perform the atomic measurements and a wide range of microwave components, sources, and existing standards.

NONINVASIVE HIGH-FREQUENCY NANOSCALE PROBING, FIELD IMAGING, AND NANOWIRE MEASUREMENTS

P. Kabos and J. R. Baker-Jarvis

Few experimental tools are currently available that can characterize and probe high-speed/frequency properties of materials and devices on a nano/molecular scale. Proposals are invited for high-frequency accurate noninvasive nanoscale probing and theory; electromagnetic field imaging; and measurements of voltage, current, and materials properties. The objective is to develop the fundamental metrology; the instruments and standards that will make it possible to perform high-speed voltage and current measurements necessary for the characterization of high-speed, high-performance ultrahigh-density nano- and molecular-electronic devices; and of the electromagnetic properties of nano-engineered, artificial, nanowire, and carbon-nanotube-based materials and devices. Film deposition, lithography and state-of-the-art microwave test systems, probe stations, and atomic force microscopy facilities are available for the proposed research.

LINEAR AND NONLINEAR DAMPING AND RELAXATION IN MAGNETIC INFORMATION STORAGE MATERIALS

P. Kabos and S. E. Russek

As magnetic recording moves well into the gigahertz frequency regime, the mechanisms that limit the precession dynamics and decay time of the spin system in magnetic thin films become limiting factors in recording head design, materials selection, and magneto-electronics applications. Low- and high-power ferromagnetic resonance (FMR), microstripline FMR, off-resonance effective linewidth techniques, and Brillouin light scattering provide critical tools needed to elucidate these processes and develop optimum thin-film materials for these applications. In this program, metallic thin materials, metal-oxide multilayers, and nano-oxide films will be fabricated and used for research on the fundamental loss processes that limit magnetic switching and large-angle precession dynamics response times. Effectively all of the established tools of high-frequency magnetics research will be used on this topic. The use of FMR techniques over a wide range of frequencies from 1 to 100 gigahertz and different static and pumping field geometries will help separate intrinsic loss processes (*e.g.*, magnon-electron scattering) from extrinsic processes (*e.g.*, two-magnon relaxation). Effective linewidth and spin-wave instability measurements will be used to probe the wave vector dependences of the intrinsic processes. High-power above-threshold spin-wave instability, Brillouin light scattering, and FMR force microscopy techniques will give new information on the actual nonlinear response of these film materials systems. This program will utilize facilities and capabilities developed at both NIST and in the laboratory of Professor Carl Patton at Colorado State University.

THERMAL NOISE MEASUREMENTS

J. Randa

Theory, standards, methods, and systems are developed for highly accurate measurements of thermal electromagnetic noise from radio frequencies to terahertz waves. Current research comprises two major thrusts: developing improved methods for characterizing the noise properties of low-noise amplifiers and transistors, and developing systems and methods for noise measurement at terahertz frequencies. Research opportunities are supported by extensive theoretical and experimental expertise, as well as world-class noise-temperature measurement capabilities.

METROLOGY FOR WIRELESS SYSTEMS

K. A. Remley

President Bush has called for broadband access for all Americans by 2007. Wireless plays an increasing role in this area. Globally, wireless revenue as a whole will approach almost 49 percent of all telecommunications services revenue (some $1.2 trillion) by the end of 2006, largely spurred by broadband. Broadband access using wireless technology requires complicated modulation formats and broad modulation bandwidths to maintain spectral efficiency. It is difficult to verify the performance of electrical measurements instrumentation for broadband wireless systems, since these instruments are currently at the cutting edge of measurement technology. In some cases, this leads to a "chicken and egg" dilemma for the wireless test industry. For example, a broadband source may be used to characterize a receiver, which is then used to characterize the source. While new hardware development will play an important role in solving the broadband measurement issue, completely new, technically sound measurement procedures and metrics are also needed to verify broadband system performance. Development of such measurement techniques will be the focus of the project. The work will provide a good mix of theoretical development and hands-on experiment in an area of intense commercial growth.

CHARACTERIZATION OF HIGH-SPEED DIGITAL INTERCONNECTS AND SIGNALS

D. F. Williams, P. Kabos, P. D. Hale, and T. S. Clement

This project provides unique opportunities for theoretical and experimental studies of high-speed digital interconnects and signals. Current research focuses on adapting microwave measurement methods to noncontacting characterization of high-speed signals and the characterization of conventional, differential, and coupled differential transmission lines fabricated on silicon substrates. This research has strong relevance to industry, and collaborators will have access to measurement equipment at frequencies beyond 110 gigahertz.

HIGH-DYNAMIC-RANGE ELECTRICAL MEASUREMENT SYSTEMS
D. F. Williams, P. D. Hale, and K. A. Remley

The ability to characterize signals that cover a high dynamic range is becoming important for many applications in both wireless and optoelectronics. Applications for high-dynamic-range measurement systems are both diverse and of high impact. They include network impairment studies for broadband wireless and optical networks, improved amplifier distortion measurements and models, detection of weak signals, and analog to digital converter characterization. We anticipate these measurement systems will also be useful in improving our traceability path for microwave phase standards. We wish to develop high-dynamic-range electrical measurement systems with measurement bandwidths up to 50 gigahertz. These new measurement systems will be based on a mixture of temporal sampling and frequency-domain down-conversion schemes. A high priority will be establishing measurement traceability to our electro-optic sampling system. This project will advance the state-of-the-art for electrical measurements and includes a good mix of both theoretical and laboratory work.

ON-WAFER MICROWAVE MEASUREMENTS AND STANDARDS
D. F. Williams and P. Kabos

Research opportunities exist for theoretical and experimental studies of wafer-level microwave measurement techniques in the frequency range of 50 megahertz to 110 gigahertz. Work focuses on the on-wafer measurement of S-parameters and signal characterization, including the development of new measurement methodologies, and modeling and analysis of single and coupled planar transmission lines. Unique opportunities exist to compare both the theoretical and experimental results.

PRECISION HIGH-SPEED WAVEFORM MEASUREMENT
D. F. Williams and P. D. Hale

We have developed an electro-optic sampling system that measures very fast electrical waveforms. This project will extend our current electrical models for the measurement system to 400 gigahertz and allow the precision measurement of waveforms with 5 ps rise times. We have constructed an electro-optic sampling system with a measurement bandwidth near one terahertz. The system is designed to accurately measure electrical pulses on transmission lines printed on special NIST-constructed electro-optic substrates, and now provides NIST's most fundamental measurements of high-speed electrical pulses. This system is now being used to calibrate the output of high-speed photodiodes and oscilloscopes to 110 gigahertz. Distortion in the probes we use to inject electrical signals onto the electro-optic substrate currently limits our measurement bandwidth to 110 gigahertz. This project will focus on using data from the electro-optic sampling system itself to characterize these probes to 200 gigahertz and beyond. The result of the effort will be models for the probes that will allow us to break our current 110 gigahertz measurement barrier and characterize the fastest photodiodes, pulse sources, and oscilloscopes in the world.

ULTRAFAST SIGNAL MEASUREMENT FOR HIGH-SPEED INTEGRATED CIRCUITS
D. F. Williams, P. Kabos, P. D. Hale, and T. S. Clement

We study methods for measuring and calibrating ultrafast signals with 110 gigahertz to 400 gigahertz bandwidths on high-speed integrated circuits. Methods that we are currently investigating include the use of electro-optic interactions with bandwidths of many terahertz. Research focuses on fully calibratable measurement of signals in printed transmission lines with low uncertainty, extension of measurements to 110 gigahertz in the near future, and extension to 400 gigahertz in the next five years.

FUNDAMENTAL MODELING FOR ELECTROMAGNETIC COMPATIBILITY
C. L. Holloway and D. A. Hill

Analytic and numerical theoretical investigations are needed on a broad range of topics related to electromagnetic compatibility (EMC) and electromagnetic interference (EMI). The particular numerical models of interest are finite-difference time-domain, finite-elements, integral equations, and hybrid techniques.

Suggested topics include printed circuit board radiation, signal integrity and coupling of high-speed digital lines and devices, lossy transmission lines, characterization and optimal design of large test facilities (*e.g.*, reverberation, anechoic, and semi-anechoic chambers), properties of electromagnetic absorbing materials, design of advanced composite and frequency selective materials, shielding effectiveness of various materials, and other coupling problems. The broad objective is to develop accurate analytic and numerical models that will advance the fundamental understanding of critical EMC/EMI issues. Extensive measurement facilities are also available with which to assess the validity of the resulting models.

INDOOR RADIO FREQUENCY PROPAGATION CHARACTERIZATION FOR BROADBAND WIRELESS: MODELING AND MEASUREMENTS

C. L. Holloway

Research for accurate characterization of general indoor propagation environments is important for the design of future wireless communications systems. It is essential to understand the efficacy of broadband wireless communications systems in office complexes and other types of building environments. The indoor radio propagation channel is a very complicated environment with a variety of propagation issues that must be defined and understood. Our research objective is to develop competence in the area of indoor and indoor-to-outdoor radio propagation, and the effects on wireless communications systems. A particular goal is to develop propagation models that will address the needs of the telecommunications industry as related to the design of state-of-the-art wireless systems that can be utilized in indoor environments. This will generally be accomplished by developing theoretical models and then designing and conducting experiments for the purposes of characterizing the indoor radio frequency (RF) propagation environment. This will lead to analytical tools that can be used by wireless system designers. Advanced computational tools, as well as excellent electromagnetic measurement facilities and instrumentation are available for experimentation. Topics that need attention include coupling of energy into building structures, propagation characterization of building materials, modeling of RF propagation into and within building structures, and measurements techniques for characterizing building environments.

THEORETICAL DEVELOPMENT OF EQUIVALENT GENERALIZED IMPEDANCE BOUNDARY CONDITIONS

C. L. Holloway

The interaction of electromagnetic fields with rough surfaces, composite materials, thin coatings, frequency selective surfaces, and particle scattering are a few of the challenging problems of current theoretical interest. Scattering problems of this type are complicated and usually require numerical techniques. However, when the detailed surface features (roughness dimension, fiber dimensions in composites, coating thickness, scattering shapes in frequency-selective surface, and particle spacing) are small compared to a wavelength, equivalent generalized impedance boundary conditions (EGIBC) can be used. These EGIBC and Maxwell's equations are all that is needed to solve these types of scattering problems. EGIBC are also very efficient in analyzing reflection problems. For example, in large electromagnetic computational codes, the use of EGIBC can eliminate the need to spatially resolve the fine detail of a particular scattering feature, which results in the abilities to solve much larger numerical problems. The proposed research direction is to use various asymptotic techniques to derive EGIBC for various electromagnetic field interactions. Specific boundary conditions will be derived so that the coefficients in the EGIBC can be interpreted in terms of electric and magnetic polarizability densities.

THEORY FOR ELECTROMAGNETIC INTERFERENCE PROBLEMS

C. L. Holloway and D. A. Hill

In spherical near-field scanning, probe pattern information and data measured at points on a spherical surface are used to determine the fields of a test antenna anywhere beyond the measurement radius (especially in the far-field zone). Although the theory is well understood, a number of areas need further work, including (1) development of more efficient measurement and computation methods; (2) analysis of and correction for measurement errors; (3) practical pattern correction schemes for general probes; (4) development of a simplified theory for the "quasi-far-field region," where measurements are "almost" made in the far field; and (5) extensions of near-field antenna theory to other applications such as acoustics, field synthesis, and the evaluation of far-field (compact range) measurement systems. We welcome proposals on these and other topics that extend or improve the application of spherical near-field measurement techniques.

ELECTROMAGNETIC THEORY FOR TIME-DOMAIN ANALYSIS

R. T. Johnk

The transient behavior of both canonical and complex objects subjected to an electromagnetic (EM) impulse from a radiating source is studied. Although transient problems can be analyzed by inverse Fourier transformation of frequency-domain data, the time-domain technique has several advantages over the frequency-domain technique. The broadband nature of time-domain analysis facilitates the understanding of nonlinearities, and the ability to model propagation delays provides improved physical insights related to the spatial geometry of the problem. Problems of both radiation and scattering from complex structures with both linear and nonlinear loadings are appropriate topics for study. The theoretical and experimental results will be applied to canonical problems to provide useful physical insights into the interaction of EM waves with devices and materials.

ELECTROMAGNETIC THEORY FOR COMPLEX ENVIRONMENTS

P. F. Wilson and D. A. Hill

Most electromagnetic field measurements seek to create a simple, well defined environment. However, real electromagnetic environments are typically complex and poorly defined. Multiple sources and scattering objects (possibly unknown or nonstationary), complicated geometries, proximity coupling, and other real world complications make electromagnetic field measurements difficult to interpret. Statistical electromagnetic approaches are needed to quantify both real environments and complex system responses. This need already exists for large systems (*e.g.*, avionics, interconnected electronics) and the requirement will move to the component level as frequencies of operation continue to move higher (*e.g.*, high-speed computers, digital wireless systems). Research opportunities exist for developing statistical electromagnetic models. Applications include reverberation chamber test methods, coupling to complex systems, shielding of ill defined geometries, radiation from statistically defined sources, propagation in nonstationary environments, and characterization of complex electromagnetic environments. The goal is to develop analytical descriptions for the statistics of these environments. Numerical modeling and Monte Carlo techniques will be used to verify analytical models. Excellent experimental facilities exist to generate measured data for comparison with theoretical and simulation results.

ANTENNA THEORY AND MEASUREMENTS I

R. C. Wittmann and M. H. Francis

A "scattering matrix description of antennas and antenna-antenna interactions" has been developed and successfully applied by researchers in this division. Radiated fields are determined from measurements made in the near field of the antenna under test, and the theory is suitable for describing antenna interactions at arbitrary distances (not just in the far-field region). Measurement techniques (and supporting theory) must continuously be extended to keep pace with the rapid advancement in antenna design and application. Topics that need attention include (1) better accuracy — high performance systems, especially those that are satellite based, require maintenance of tighter tolerances; (2) higher frequencies — more sophisticated near-field measurement methods are needed to handle millimeter-wave applications (up to above 500 gigahertz); (3) complex phased-array antennas — large, often electronically steerable, phased arrays need special diagnostic tests to ensure optimum functionality; (4) low sidelobe antennas — military and commercial communication applications increasingly specify sidelobe levels 50 dB or more below peak, a range where measurement by standard technique is difficult; (5) *in situ* (or remote) evaluation — many systems cannot be transported easily to a measurement laboratory; robust methods are needed for on-site testing; and (6) multiple reflections — methods are needed to mitigate errors caused by multiple reflections between the probe and test antenna. Research facilities include two planar near-field scanners, a multipurpose range for cylindrical and spherical antenna measurements, a precision extrapolation range, an anechoic chamber, a ground screen, and other EM experimental facilities, as well as excellent computational resources. We welcome proposals in these or related areas that extend or improve the application of near-field antenna measurement methods.

ANTENNA THEORY AND MEASUREMENTS II

R. C. Wittmann and M. H. Francis

In spherical near-field scanning, probe pattern information and data measured at points on a spherical surface are used to determine the fields of a test antenna anywhere beyond the measurement radius (especially in the far-field zone). Although the theory is well understood, a number of areas need further work, including (1) development of more efficient measurement and computation methods; (2) analysis of and correction for measurement errors; (3) practical pattern correction schemes for general probes; (4) development of a simplified theory for the "quasi-far-field region" where measurements are "almost" made in the far field; and (5) extensions of near-field antenna theory to other applications such as acoustics, field synthesis, and the evaluation of far-field (compact range) measurement systems. We welcome proposals on these and other topics that extend or improve the application of spherical near-field measurement techniques.

HIGH-FIELD SUPERCONDUCTOR RESEARCH

J. W. Ekin

This research program is interdisciplinary, encompassing the physical, mechanical, and electrical properties of high-field superconducting materials and composites. Research is conducted on new types of high current density superconductors and superconductor stabilization that are being developed for high energy physics accelerators, magnetic fusion systems, and magnetic resonance systems. Experimental programs include the study of the electromechanical properties of superconductors, high-pinning materials, internal reinforcement, and advance composite design. Very high-field magnet systems, power supplies, servo-hydraulic mechanical test systems, and analytic microscopy facilities are available. Theoretical studies concentrate on flux pinning and the intrinsic effect of strain on the superconducting state.

HIGH T_c SUPERCONDUCTOR RESEARCH

J. W. Ekin

A unique feature of this program is that we use our expertise in electromechanical measurements and contact interface studies to work more closely with research laboratories in industry and at major universities to develop new types of high-T_c superconductors for emerging power technologies in transmission lines and high-power-density rotating machinery. We are interested in all aspects of high-T_c superconductor research, including the mechanical, magnetic field, and electrical limits of the new class of thick-film YBCO coated conductors. Our film deposition and cleanroom facilities are fully equipped for fabrication and micropatterning of film conductors, contacts, and research test structures. Particular effort now focuses on understanding the basic properties of YBCO film superconductors, new reinforced low porosity Bi-2223 and Bi-2212 superconductors, and MgB_2 multifilamentary composite conductors.

FERRITIN AND OTHER MAGNETIC NANOPARTICLES IN PROTEIN SHELLS

R. B. Goldfarb

Ferritin is nature's ubiquitous iron-storage molecule, found in species ranging from microbes to man. It consists of a roughly spherical apoferritin protein shell, inside which iron accumulates in the form of a ferric oxyhydroxide crystal. The outer diameter is 12 nm, irrespective of the amount of iron stored within. Although its physical, chemical, and magnetic properties have been studied for more than 60 years, ferritin remains a subject of current research, with many implications for biology and medicine. In particular, ferritin is an important contributor to T_1 and T_2 relaxation, which effectively determine image contrast in magnetic resonance imaging. Applicant with backgrounds in physical chemistry, chemical synthesis, magneto-chemistry, magnetometry, nuclear magnetic resonance, and electron paramagnetic resonance are invited to apply. Equipment available includes a 7-tesla SQUID magnetometer and a 7-tesla/300-megahertz nuclear magnetic resonance system.

Metrology for Magnetic Resonance Imaging Artifacts and Stability

L. F. Goodrich and S. E. Russek

Research focuses on developing metrology for magnetic resonance imaging (MRI) system stability and intercomparability and image distortion. This is part of a new NIST effort to assist the medical imaging community in developing more quantitative images that are traceable to the International System of Units (SI). We intend to conduct studies to assess MRI system stability and intercomparability using NIST-generated artifacts and assess methods to identify sources of instability in MRI coils systems: superconducting solenoid, shim coils, gradient coils, and RF coils. We plan to develop metrology to analyze magnetic imaging distortions as a result of implants such as stents and coils. This will involve evaluating magnetic properties of MRI system components and implants that could lead to image distortion. Measurements will be made using a 7 tesla SQUID magnetometer, a 7-tesla/300-megahertz nuclear magnetic resonance system, and other superconducting high-magnetic field systems at NIST-Boulder. All of these studies will also involve measurements conducted in commercial MRI systems at local hospitals and at the National Institutes of Health (NIH). This opportunity is appropriate for physicists, chemical-physicists, or engineers interested in medical physics or medical engineering.

Superconductor Measurements

L. F. Goodrich

We develop and evaluate measurement techniques to determine the critical parameters and matrix properties of superconductors. Capabilities include variable-temperature critical-current measurement, low-noise current supplies up to 3000 amperes, high-field magnets up to 18 teslas, and voltage sensitivity to 1 nanovolt. We study conventional superconductors (NbTi and Nb_3Sn) and the newer high-transition-temperature materials. We conduct fundamental studies of the superconducting-normal transitions and the parameters that affect their accurate determination, such as current transfer, strain, or inhomogeneities in materials and fields. We develop theoretical models to interpret current redistribution and component interactions in composite superconductors.

Cellular Imaging and Manipulation with Superparamagnetic Particles (Joint NIH/NIST)

J. Moreland

We are currently developing technologies for characterizing and manipulating magnetic particles for *in-vivo* magnetic resonance imaging (MRI) and cell manipulation. Applications include tags and contrast agents for imaging single cells in organs as well as very high resolution imaging of the organelles in a single cell. In addition, we are developing ways of using magnetic particles to move to predetermined locations or to switch cell functionality. Understanding the nanoscopic physical properties of the particles is critical for exploring and tailoring them to specific applications. The main areas of interest include: (1) characterizing the size, shape, biofunctionality, and nanoscopic field patterns of superparamagnetic particles with the objective of developing methods for narrowing the statistical distributions of each of these properties; (2) understanding natural biological magnetic systems and their physicochemical processes; and (3) developing magnetic force manipulation techniques for moving and switching cell functionality and for injecting cells with superparamagnetic particles.

Chip-Scale Magnetic Resonance Imaging Microscope

J. Moreland

Conventional magnetic resonance imaging (MRI) systems are limited in resolution because of the noise of inductive detectors, which limits sensitivity, as well as difficulties generating field gradients sufficient for narrow sample slice discrimination. By using micromechanical cantilever oscillators as magnetic sensors and by reducing the dimensions of the gradient coils, significant improvements in sensitivity and resolution can be made. In particular, we have recently demonstrated several micromechanical magnetometers including a magnetic resonance force microscope, a torque magnetometer, and a micromechanical calorimeter. These magnetometers operate on the principle of modulating the magnetic resonance excitation of a sample attached to a microcantilever at the cantilever's resonance frequency. Our goal is to optimize these novel detectors for biological applications, including in-vivo imaging of cell organelles and membranes. The main challenge is to develop integrated microsystems with microfabricated DC and RF field sources, magnetic sensors, and field gradient coils. Ideally, the chip should be adaptable to microfluidic environments.

MICRO-ELECTROMECHANICAL SYSTEMS FOR METROLOGY

J. Moreland

We are developing micro-electromechanical systems (MEMS) with integrated components for precision measurement purposes. Work focuses on the following goals: (1) improving the performance of fundamental standards instrumentation by developing novel detectors and more fully integrated measurement systems, (2) exploring the impact of MEMS and MEMS-based metrology on the future development of the microelectronics and data storage industries, and (3) improving the manufacturing yield with MEMS probe assemblies designed for production line testing. Our cleanroom facility is fully equipped for bulk and surface micromachining of silicon wafers including design, fabrication, and testing tools. We are interested in all aspects of research including the development of novel MEMS structures, as well as the testing and integration of MEMS structures into precision measurement instruments.

NANOSCALE IMAGING FOR MAGNETIC TECHNOLOGY

J. Moreland

The magnetic storage industry has advanced to the stage where nanometer-scale morphological and physical properties play an important role in current and future disk drive performance. In its many forms, scanned probe microscopy (SPM) can be used to measure roughness, device dimensions, electromagnetic field patterns, and various physical processes at nanometer scales, which provides important information about the fundamental operation and limitations of drive components. Our goal is to help tailor SPM techniques for these applications. We are investigating scanning tunneling microscopy, atomic force microscopy, magnetic force microscopy, scanning potentiometry, and scanning thermometry for their usefulness.

SINGLE-MOLECULE MANIPULATION AND MEASUREMENT

J. Moreland

We are developing a nano-electromechanical system platform for single biomolecule manipulation and measurement. Measurements to determine the structure and function of protein and DNA are currently made using large populations of molecules rather than single molecules. Researchers in biotechnology have shown that the behavior of single molecules in living systems can be different from results obtained by measuring the statistical average of large populations of molecules. The limitation in making single molecule measurements is primarily due to the lack of measurement tools and methods that are capable of isolating, manipulating, and probing the behavior and structure of the molecules. As a result, there is a rapidly growing interest in the development and application of nanotechnology to support single-molecule measurements.

SPIN ELECTRONICS

W. H. Rippard, S. E. Russek, and T. J. Silva

Until recently, the only means known to control the magnetization state of ferromagnetic structures was through the use of applied magnetic fields. However, within the last several years it has been demonstrated that this can also be accomplished through the transfer of the electron spin angular momentum from current-carrying electrons to the magnetization of magnetic films, generally referred to as the spin-momentum-transfer effect. Spin transfer represents a fundamentally new way to control and manipulate the magnetic states of devices, and allows hysteretic switching and coherent microwave dynamics to be excited in magnetic nanostructures using a DC current. This project seeks both to understand the fundamental characteristics of the interaction between spin polarized currents and magnetic materials, and also to examine the suitability of such nanoscale devices for microwave electronics. We are specifically pursuing research in: (1) increasing output power from nanoscale oscillators through materials engineering and incorporating tunnel junctions into the device structures, (2) understanding the interactions between mutually synchronized nanoscale oscillators in order to develop device arrays, (3) characterizing and understanding the thermal contributions to both oscillator linewidths and the current induced switching distributions in patterned elements, (4) understanding the interactions between individual magnetic nanostructures and AC fields and AC currents, and (5) investigating the current-induced switching properties of patterned magnetic nanostructures for magnetic random-access memory (MRAM) applications.

Magnetic Sensors, Spintronic Materials, and Nanomagnetic Imaging

S. E. Russek and W. H. Rippard

Spin-dependent transport is a widely used, yet poorly understood, phenomenon. Giant magnetoresistive (GMR) devices and magnetic tunnel junctions (MTJ) are being developed for use in magnetic recording heads, magnetoresistive random access memories (MRAM), and magnetic sensors for industrial and biomedical applications. The goal of this research opportunity is to develop a better fundamental understanding of spin-dependent transport in magnetic metals, normal metals, conducting oxides, and semiconductors. Research areas include understanding spin polarization in exotic materials, examining the effects of interfaces, and looking at the nanomagnetic structure that gives rise to noise and device-to-device variation. This research involves fabrication of novel magnetic devices such as GMR, MTJ, and spin-transfer devices using a state-of-the-art, eight-source, ultrahigh vacuum deposition system and a combination of optical, electron-beam, and scanned-probe lithography. Electrical measurements, over a wide range of temperature, field, and frequency, will be used to characterize spin-dependent transport properties with a particular emphasis on high-frequency and noise properties of the devices. Lorentz and magnetic force microscopy will be used to characterize nanoscale magnetic structures. New types of dynamical nanoscale imaging, such as those based on ballistic electron magnetic microscopy (BEMM), will be developed.

Biomagnetic Imaging

S. E. Russek and P. Kabos

This research will focus on developing advanced magnetic resonance imaging (MRI). Nanomagnetic contrast agents will be fabricated and characterized. The nanomagnets to be investigated include molecular nanomagnets, such as Fe_8, and nanoparticles formed by novel thin-film self assembly techniques. The nanomagnets will be characterized using SQUID magnetometry and electron spin resonance (ESR) at fields up to 7 teslas and frequencies up to 140 gigahertz. The affect of the nanomagnets on nuclear magnetic resonance (NMR) properties of protons in biological solutions will be studied. Systems that can simultaneously measure NMR, ESR, and nanomagnet magnetization will be developed. A detailed understanding of how the nanomagnet properties, such as anisotropies and fluctuations, affect proton relaxation in biological solutions will be developed. MRI imaging of potential nanomagnetic contrast agents will be done on commercial MRI systems at local hospitals. Materials with different relaxation times will be investigated for use in MRI standards and phantoms.

High-Frequency Characterization of Novel Thin-Film Materials

S. E. Russek, J. R. Baker-Jarvis, and P. Kabos

The goal of this project is to fabricate novel nano-engineered thin-film materials and measure their electromagnetic properties in the 1 to 100 gigahertz regime. The materials include nanostructured materials, composite ferromagnetic-ferroelectric materials, "left-handed" materials, and frequency-tunable materials. The materials can be fabricated with an ultrahigh vacuum, eight-source sputtering system, a laser ablation system, and optical and electron-beam lithography systems. The dielectric and magnetic properties can be engineered by patterning arrays of elements on two different length scales. Patterning on a scale comparable to the excitation wavelength — about 1 millimeter — will allow the development of artificial crystals (photonic band gap materials) in the microwave regime. Patterning on a scale much shorter than the wavelength, 10 to 100 nm, will allow the permittivity, permeability, and conductivity to be engineered and controlled to have new functionalities. Examples of such materials engineering include light- and field-tunable exchange coupling, low loss amorphous/nanoparticle composites, negative-epsilon negative-mu ("left handed") systems, and ferroelectric-ferromagnetic multilayers. Measurements will be conducted on state-of-the-art, 100 gigahertz microwave test systems and cryogenic microwave probe stations.

MOLECULAR AND NANOMAGNETISM

S. E. Russek

Ultrasmall magnetic structures will be fabricated using both conventional nano-lithography techniques (electron-beam and scanned-probe lithographies) and chemical synthetic techniques. The systems studied may include molecular nanomagnets (*e.g.*, Mn-12), carbon nanotubes grown on magnetic nanoparticles, patterned longitudinal and perpendicular media, and nanodevices. The goal of this research will be to understand the physics of ultrasmall magnetic structures, their implications for the limits of magnetic data storage, and to develop novel nanoscale devices. The magnetization and switching processes will be studied as a function of size, shape, and temperature to characterize thermally activated and quantum mechanical tunneling transitions. The high-frequency properties (1 gigahertz to 150 gigahertz) will be studied with high-frequency electron spin resonance, ferromagnetic resonance, and transport properties. The molecular nanomagnets and magnetic nanostructures will be incorporated into thin-film device structures to explore potential device applications.

HIGH-SPEED MAGNETIC PHENOMENA

T. J. Silva

Experimental methods to determine fundamental limits to the data transfer rate of magnetic devices are being developed. Both low-coercivity ("soft") and high-coercivity ("hard") magnetic materials are studied. Experimental techniques include electrically sampled inductive detection and time-resolved magneto-optics for the study of soft magnetic materials. Quantitative Kerr microscopy is used for the measurement of switching speed in hard magnetic materials. Extensive facilities include a 20-gigahertz sampling oscilloscope, a 50-femtosecond mode-locked Ti:sapphire laser, and a digital Kerr microscope with a high-performance chilled charge-coupled device camera. Commercial and experimental solid-state instrumentation is used for the generation of microwave pulses. Waveguide technology is employed to deliver subnanosecond magnetic field pulses to samples. Waveguide structures are lithographically fabricated on site in a state-of-the-art cleanroom, which includes mask generation facilities. Applicants are encouraged who have a strong experimental background in magnetism, especially high-frequency magnetic phenomena such as ferromagnetic resonance.

NONLINEAR MAGNETO-OPTICS

T. J. Silva

The second-harmonic magneto-optic Kerr effect (SHMOKE) is under investigation as a tool for the study of interfacial magnetism. SHMOKE shows strong sensitivity to the magnetization at optically accessible interfaces between ferromagnetic and non-ferromagnetic films, yet SHMOKE does not require exotic facilities, such as ultrahigh vacuum (UHV) or synchrotron radiation. Therefore, SHMOKE shows great promise as an industrial diagnostic instrument for the optimization of giant magnetoresistive sensors and magnetic tunnel junctions, where interfacial magnetism strongly influences device performance. SHMOKE also exhibits a strong magneto-optic signal, with the magnetic contrast approaching 60 percent in some sample systems. Extensive resources for the study of SHMOKE include a mode-locked 50-femtosecond Ti:sapphire laser, coincident photon detection electronics, photo-elastic modulators, lock-in amplifiers, and sample translation stages. Samples may be produced on site with a state-of-the-art, eight-source UHV sputtering system. Applicants are preferred with a strong experimental background in magnetic thin films, magnetic multilayers, magneto-optics, and/or nonlinear optics.

Thermal Instability of Magnetic Thin Films

T. J. Silva

As the grain size of thin-film magnetic recording media steadily decreases with increasing areal capacities, we are concerned that recorded information may be erased as a result of thermally activated switching of the individual grains — the so-called "superparamagnetic limit." Our goal is to understand the fundamental mechanisms that result in thermal erasure through the measurement of various phenomena, including magnetic viscosity and the time dependence of coercivity. Emphasis is placed on determining the thermal stability of media over a wide range of time scales, from those accessible with large-scale magnetometers to those that use pulsed microwave fields. The final goal is a measurement technique for the determination of data stability in media without resorting to mean-time-before-failure analysis. Extensive facilities include numerous magnetometers (vibrating-sample magnetometer, alternating gradient magnetometer, SQUID magnetometer), a transmission electron microscope for the determination of grain size, and a state-of-the-art, eight-source, ultrahigh-vacuum sputtering system for the preparation of samples. Applicants with a strong experimental background in magnetism — especially magnetic thin-film preparation and characterization — are encouraged to apply.

Appendix C:
Prefixes for the International System of Units (SI)

Multiplication Factor	Prefix	Symbol	Multiplication Factor	Prefix	Symbol
10^{24}	yotta	Y	10^{-1}	deci	d
10^{21}	zetta	Z	10^{-2}	centi	c
10^{18}	exa	E	10^{-3}	milli	m
10^{15}	peta	P	10^{-6}	micro	µ
10^{12}	tera	T	10^{-9}	nano	n
10^{9}	giga	G	10^{-12}	pico	p
10^{6}	mega	M	10^{-15}	femto	f
10^{3}	kilo	k	10^{-18}	atto	a
10^{2}	hecto	h	10^{-21}	zepto	z
10^{1}	deka	da	10^{-24}	yocto	y

Appendix D:
Units for Magnetic Properties

Symbol	Quantity	Conversion from Gaussian and cgs emu to SI
Φ	magnetic flux	$1\ Mx \rightarrow 10^{-8}\ Wb = 10^{-8}\ V \cdot s$
B	magnetic flux density, magnetic induction	$1\ G \rightarrow 10^{-4}\ T = 10^{-4}\ Wb/m^2$
H	magnetic field strength	$1\ Oe \rightarrow 10^3/(4\pi)\ A/m$
m	magnetic moment	$1\ erg/G = 1\ emu \rightarrow 10^{-3}\ A \cdot m^2 = 10^{-3}\ J/T$
M	magnetization	$1\ erg/(G \cdot cm^3) = 1\ emu/cm^3 \rightarrow 10^3\ A/m$
$4\pi M$	magnetization	$1\ G \rightarrow 10^3/(4\pi)\ A/m$
σ	mass magnetization, specific magnetization	$1\ erg/(G \cdot g) = 1\ emu/g \rightarrow 1\ A \cdot m^2/kg$
j	magnetic dipole moment	$1\ erg/G = 1\ emu \rightarrow 4\pi \times 10^{-10}\ Wb \cdot m$
J	magnetic polarization	$1\ erg/(G \cdot cm^3) = 1\ emu/cm^3 \rightarrow 4\pi \times 10^{-4}\ T$
χ, κ	susceptibility	$1 \rightarrow 4\pi$
χ_ρ	mass susceptibility	$1\ cm^3/g \rightarrow 4\pi \times 10^{-3}\ m^3/kg$
μ	permeability	$1 \rightarrow 4\pi \times 10^{-7}\ H/m = 4\pi \times 10^{-7}\ Wb/(A \cdot m)$
μ_r	relative permeability	$\mu \rightarrow \mu_r$
w, W	energy density	$1\ erg/cm^3 \rightarrow 10^{-1}\ J/m^3$
N, D	demagnetizing factor	$1 \rightarrow 1/(4\pi)$

Gaussian units are the same as cgs emu for magnetostatics; Mx = maxwell, G = gauss, Oe = oersted, Wb = weber, V = volt, s = second, T = tesla, m = meter, A = ampere, J = joule, kg = kilogram, H = henry.

APPENDIX E:
SYMBOLS FOR THE CHEMICAL ELEMENTS

Symbol	Element	Symbol	Element	Symbol	Element
Ac	Actinium	Gd	Gadolinium	Po	Polonium
Ag	Silver	Ge	Germanium	Pr	Praseodymium
Al	Aluminum	H	Hydrogen	Pt	Platinum
Am	Americium	He	Helium	Pu	Plutonium
Ar	Argon	Hf	Hafnium	Ra	Radium
As	Arsenic	Hg	Mercury	Rb	Rubidium
At	Astatine	Ho	Holmium	Re	Rhenium
Au	Gold	I	Iodine	Rh	Rhodium
B	Boron	In	Indium	Rn	Radon
Ba	Barium	Ir	Iridium	Ru	Ruthenium
Be	Beryllium	K	Potassium	S	Sulfur
Bi	Bismuth	Kr	Krypton	Sb	Antimony
Bk	Berkelium	La	Lanthanum	Sc	Scandium
Br	Bromine	Li	Lithium	Se	Selenium
C	Carbon	Lr	Lawrencium	Si	Silicon
Ca	Calcium	Lu	Lutetium	Sm	Samarium
Cd	Cadmium	Md	Mendelevium	Sn	Tin
Ce	Cerium	Mg	Magnesium	Sr	Strontium
Cf	Californium	Mn	Manganese	Ta	Tantalum
Cl	Chlorine	Mo	Molybdenum	Tb	Terbium
Cm	Curium	N	Nitrogen	Tc	Technetium
Co	Cobalt	Na	Sodium	Te	Tellurium
Cr	Chromium	Nb	Niobium	Th	Thorium
Cs	Cesium	Nd	Neodymium	Ti	Titanium
Cu	Copper	Ne	Neon	Tl	Thallium
Dy	Dysprosium	Ni	Nickel	Tm	Thulium
Er	Erbium	No	Nobelium	U	Uranium
Es	Einsteinium	Np	Neptunium	V	Vanadium
Eu	Europium	O	Oxygen	W	Tungsten
F	Fluorine	Os	Osmium	Xe	Xenon
Fe	Iron	P	Phosphorus	Y	Yttrium
Fm	Fermium	Pa	Protactinium	Yb	Ytterbium
Fr	Francium	Pb	Lead	Zn	Zinc
Ga	Gallium	Pd	Palladium	Zr	Zirconium
		Pm	Promethium		

Division Publication Editor:	Ron Goldfarb
Publication Coordinator:	Erik M. Secula
Printing Coordinator:	Ilse Putman
Document Production:	Technology & Management Services, Inc. Gaithersburg, Maryland

January 2007

For additional information contact:
Telephone: (303) 497-3131
Facsimile: (303) 497-3122
On the Web: http://www.boulder.nist.gov/div818/